PROCEEDINGS OF THE 11th ASIAN LOGIC CONFERENCE

In Honor of Professor Chong Chitat on His 60th Birthday

PROCEEDINGS OF THE
11th ASIAN LOGIC CONFERENCE
In Honor of Professor Chong Chitat on His 60th Birthday

National University of Singapore, Singapore   22 – 27 June 2009

Edited by

Toshiyasu Arai
Chiba University, Japan

Qi Feng
National University of Singapore, Singapore

Byunghan Kim
Yonsei University, South Korea

Guohua Wu
Nanyang Technological University, Singapore

Yue Yang
National University of Singapore, Singapore

NEW JERSEY • LONDON • SINGAPORE • BEIJING • SHANGHAI • HONG KONG • TAIPEI • CHENNAI

Published by

World Scientific Publishing Co. Pte. Ltd.
5 Toh Tuck Link, Singapore 596224
USA office: 27 Warren Street, Suite 401-402, Hackensack, NJ 07601
UK office: 57 Shelton Street, Covent Garden, London WC2H 9HE

Library of Congress Cataloging-in-Publication Data
Asian Logic Conference (11th : 2009 : Singapore)
Proceedings of the 11th Asian Logic Conference : in honor of Professor Chong Chitat on his 60th birthday, National University of Singapore, Singapore, 22–27 June 2009 / edited by Toshiyasu Arai ... [et al.].
p. cm.
Includes bibliographical references and index.
ISBN-13: 978-981-4360-53-1 (hardcover : alk. paper)
ISBN-10: 981-4360-53-8 (hardcover : alk. paper)
1. Logic, Symbolic and mathematical--Congresses. I. Chong, C.-T. (Chi-Tat), 1949–
II. Arai, T. (Toshiyasu) III. Title.
Q334.A853 2009
511.3--dc23

2011020566

British Library Cataloguing-in-Publication Data
A catalogue record for this book is available from the British Library.

Printed in Singapore.

PREFACE

The 11th Asian Logic Conference was held June 22 - 27, 2009 at National University of Singapore, Singapore. The Asian Logic Conferences are major international events in mathematical logic. From 1981 to 2008, they had been held triennially and rotated among countries in the Asia-Pacific region. In 2008, the East Asian and Australasian Committees of the Association of Symbolic Logic decided to shorten the three-year cycle to two. The committees also chose Singapore 2009 to begin the new cycle in honor of Professor Chitat Chong's 60th birthday. Professor Chong is one of the founders of the Asian Logic Conference series and a central figure in establishing mathematical logic in Asia.

Major funding was provided by Faculty of Science, Institute for Mathematical Sciences, and Office of the Provost of National University of Singapore; the Association of Symbolic Logic, and World Scientific Publishing Company.

The program consisted of 17 one-hour plenary lectures given by Klaus Ambos-Spies, Toshiyasu Arai, Bektur Baizhanov, John T. Baldwin, Rodney Downey, Ilijas Farah, Renling Jin, Iskander Sh. Kalimullin, Peter Koellner, Manuel Lerman, Menachem Magidor, Michael Rathjen, Gerald E. Sacks, Stephen G. Simpson, Theodore A. Slaman, Frank Stephan, W. Hugh Woodin; and two parallel sessions of 32 contributed talks (30 minutes each).

There were 85 registered participants from the following countries and regions: Canada, China, Germany, India, Israel, Japan, Kazakhstan, New Zealand, Poland, Russia, South Korea, Spain, Taiwan, United Kingdom, United States of America, and Singapore.

Participants were invited to submit papers to the present volume. All of the submitted papers were fully refereed.

We wish to thank all those who help to review the submitted papers.

Editors

Toshiyasu Arai, Qi Feng, Byunghan Kim, Guohua Wu and Yue Yang

ORGANIZING COMMITTEES

Program Committee

Rodney Downey – Victoria University of Wellington, New Zealand
Qi Feng (Chair) – National University of Singapore, Singapore and Chinese Academy of Sciences, China
Byunghan Kim – Yonsei University, South Korea
Theodore A. Slaman – University of California at Berkeley, USA
Akito Tsuboi – University of Tsukuba, Japan
W. Hugh Woodin – University of California at Berkeley, USA
Yue Yang – National University of Singapore, Singapore

Local Organizing Committee

Qi Feng – National University of Singapore, Singapore and Chinese Academy of Sciences, China
Frank Stephan – National University of Singapore, Singapore
Guohua Wu – Nanyang Technological University, Singapore
Yue Yang (Chair) – National University of Singapore, Singapore

CONTENTS

PROVABLY Δ_2^0 AND WEAKLY DESCENDING CHAINS*

Toshiyasu Arai

Graduate School of Science, Chiba University
1-33, Yayoi-cho, Inage-ku, Chiba, 263-8522, Japan
tosarai@faculty.chiba-u.jp

In this note we show that a set is provably Δ_2^0 in the fragment $I\Sigma_n$ of arithmetic iff it is $I\Sigma_n$-provably in the class D_α of α-r.e. sets in the Ershov hierarchy for an $\alpha <_{\varepsilon_0} \omega_{1+n}$, where $<_{\varepsilon_0}$ denotes a standard ε_0-ordering.

In the Appendix it is shown that a limit existence rule $(LimR)$ due to Beklemishev and Visser becomes stronger when the number of nested applications of the inference rule grows.

1. Introduction

Thoroughout this paper, we identify a predicate A with its characteristic function

$$A(x_1, \dots, x_n) = \begin{cases} 0 \text{ if } A(x_1, \dots, x_n) \\ 1 \text{ otherwise} \end{cases}$$

Natural numbers c are identified with the sets $\{n \in \mathbb{N} : n < c\}$.

The following Limit Lemma due to Shoenfield is a classic in computability theory.

Theorem 1 (Limit Lemma)
A set A of natural numbers is Δ_2^0 iff there is a binary (primitive) recursive predicate $f : \omega \times \omega \to 2 = \{0, 1\}$ such that

$$\forall c[\lim_{w\to\infty} f(c, w) = A(c)].$$

Moreover the theorem is provable uniformly in $B\Sigma_1^0 \subsetneq I\Sigma_1^0$, cf. [4], pp. 89-91. Let us call the predicate f a *witnessing predicate* for $A \in \Delta_2^0$.

In this paper we address a problem asking what can we say about the rate of convergences of the predicate f under the assumption that the set A is *provably* Δ_2^0 in a formal (sound) theory T?

*Dedicated to the occasion of Chong Chi Tat's 60th birthday.

This is a problem on a hierarchy. The class of Δ^0_2-sets is classified in the Ershov hierarchy, [3]. A recent article [11] due to F. Stephan, Y. Yang and L. Yu is a readable contribution to the hierarchy, to which we refer as a standard text.

The α-th level of the Ershov hierarchy is denoted D_α for notations α of constructive ordinals, and a set in D_α is said to be an α-r.e. set.

It is known, as usual in hierarchic problems indexed by constructive ordinals, that D_α depends heavily on notations α, i.e., the order type of α does not determine the set D_α. By reason of this dependency let us fix a standard elementary recursive well ordering $<_\alpha$ of type α. I don't want to discuss here what is a 'standard ordering' or a 'natural well ordering'. We assume that $\mathsf{EA}=\mathrm{I}\Delta^0_0+exp$, Elementary Recursive Arithmetic, proves some algebraic facts on the ordering $<_\alpha$. For the case $\alpha=\varepsilon_0$, what we need on $<_{\varepsilon_0}$ can be found in, e.g., [10].

In what follows let us drop the subscript α in $<_\alpha$ when no confusion likely occurs.

Definition 2 (Stephan-Yang-Yu [11])
Let $K \in dom(<)$, the domain of the order $<$.

A set A of natural numbers is *K-r.e. with respect to* $<$ iff there exist a binary *recursive* predicate f, and a *recursive* function $h : \omega \times \omega \to K = \{\beta \in dom(<) : \beta < K\}$ *such that*

1.

$$\text{(weakly descending) } K > h(c,w) \geq h(c,w+1) \tag{1}$$

2.

$$\text{(lowering) } f(c,w) \neq f(c,w+1) \to h(c,w) > h(c,w+1) \tag{2}$$

3.

$$\forall c[\lim_{w\to\infty} f(c,w) = A(c)] \tag{3}$$

Roughly speaking, a set is K-r.e. if the convergence of its witnessing predicate follows from the fact that weakly decreasing functions in K have to be constant eventually.

Now suppose that we have a proof-theoretic analysis of a formal (and sound) theory T, e.g., a cut-elimination through a transfinite induction along a standard well ordering $<$. It, then, turns out that A is provably Δ^0_2 in T iff T proves the fact that $A \in D_K$ with respect to $<$ for a $K \in dom(<)$.

Though, in this paper, we restrict our attention to $\mathrm{T} = I\Sigma^0_n$ of fragments of first order arithmetic as a concrete example, where the order $<$ denotes a standard well ordering of type ε_0, it is easy to see that our proof works also for stronger theories, e.g., second order arithmetic Π^1_1-CA_0 and fragments of set theories.

In Section 2 it is shown that for each $n \geq 1$, a set is provably Δ^0_2 in the fragment $I\Sigma_n$ iff it is $I\Sigma_n$-provably in the class D_α for an $\alpha <_{\varepsilon_0} \omega_{1+n}$ (Theorem 4).

Also any provably Σ^0_2-function has a Skolem function $F(c) = \lim_{w\to\infty} f(c, w)$ as limits of an f, whose convergence is ensured by weakly descending chains of ordinals (Theorem 9). Moreover the 2-consistency $\mathrm{RFN}_{\Pi^0_3}(I\Sigma^0_n)$ is seen to be equivalent over Primitive Recursive Arithmetic PRA to the fact that every primitive recursive weakly descending chain of ordinals$< \omega_{1+n}$ has a limit (Theorem 10).

In Section 3 it is shown that a set is provably Δ^0_2 in Elementary Recursive Arithmetic EA iff it is EA-provably in the class D_n of a finite level (Theorem 11). Our proof seems to be a neat application of the Herbrand's theorem.

The Appendix A contains another application of Herbrand's theorem. We consider, over EA, an inference rule $(LimR)$ in [2], which concludes the convergence of an elementary recursive series $\{h(n)\}_n$ under the assumption that the series is weakly decreasing almost all n. Note that $(LimR)$ is an inference rule, and not an axiom(sentence).

On the other side, let $L\Sigma_1^{-(k)}$ denote the schema in [5], saying that any non-empty Σ^0_1 k-ary predicate has the least tuple, which is least with respect to the lexicographic ordering on $\mathbb{N}^k$.

It is shown that $L\Sigma_1^{-(k)}$ is equivalent to the k-nested applications of $(LimR)$. In [5], Corollary 2.11 it was shown that $\{L\Sigma_1^{-(k)}\}_k$ forms a proper hierarchy, i.e., $L\Sigma_1^{-(k+1)} \vdash \mathrm{Con}(L\Sigma_1^{-(k)})$. Hence we conclude that a $(k+1)$-nested application of $(LimR)$ proves the consistency of the k-nested applications of $(LimR)$.

2. Provably Δ^0_2 in $I\Sigma^0_n$

Let LEA [EA] denote the Lower Elementary Recursive Arithmetic [Elementary Recursive Arithmetic], which is a first-order theory in the language having function constants for each code(algorithm) of lower elementary recursive function [function constants for each code of elementary recursive function], resp. Cf. [8] and [9] for these classes of subrecursive functions.

Induction schema is restricted to quantifier-free formulas in the language. The axioms of the theories LEA, EA are purely universal ones.

Let $I\Sigma^0_n$ denote the fragment of arithmetic, which is a first-order theory in the language of LEA, and Induction schema is restricted to Σ^0_n formulas. Here a Σ^0_0 formula is a quantifier-free formula. $I\Sigma^0_0$ is another name for LEA.

Let $<_{\varepsilon_0}$ denote a standard ε_0-ordering. We assume that EA proves some algebraic facts on the ordering $<_{\varepsilon_0}$. What we need on $<_{\varepsilon_0}$ can be found in, e.g., [10].

In what follows let us drop the subscript ε_0 in $<_{\varepsilon_0}$ when no confusion likely occurs.

For a class Φ of formulas and an ordinal α let $TI(\Phi, \alpha)$ denote the schema of transfinite induction up to α and applied to a formula $\varphi \in \Phi$:

$$\forall \beta[\forall \gamma < \beta \varphi(\gamma) \to \varphi(\beta)] \to \forall \beta < \alpha \varphi(\beta).$$

Let

$$\omega_0 := 1, \; \omega_{1+n} := \omega^{\omega_n}.$$

Here is a folklore result on provability of the restricted transfinite induction schemata in fragments of arithmetic.

Theorem 3 (See, e.g., [10])
For each $n \geq 0$, $I\Sigma^0_n \vdash TI(\Pi^0_1, \alpha)$ iff $\alpha < \omega_{1+n}$.

The following Theorem 4 states that for positive integers n, a set is provably Δ^0_2 in $I\Sigma^0_n$ iff it is $I\Sigma^0_n$-provably in the class D_α of α-r.e. sets in the Ershov hierarchy for an $\alpha <_{\varepsilon_0} \omega_{1+n}$. Moreover (weakly descending) and (lowering) are provable in EA.

Theorem 4 *For positive integers n, the following are equivalent for quantifier-free A, B and a free variable c.*

1. *$I\Sigma^0_n$ proves*

$$\forall x \exists y A(x, y, c) \leftrightarrow \exists z \forall u B(z, u, c) \tag{4}$$

2. *There exists a binary* elementary *recursive predicate f, an ordinal $K < \omega_{1+n}$ and an* elementary *recursive function $h : \omega \times \omega \to K$ such that*

   (a) (weakly descending)

$$\mathsf{EA} \vdash K > h(c, w) \geq h(c, w + 1)$$

(b) (lowering)

$$\mathsf{EA} \vdash f(c,w) \neq f(c,w+1) \to h(c,w) > h(c,w+1)$$

(c)

$$\mathsf{EA} \vdash \lim_{w\to\infty} f(c,w) = 0 \to \exists z \forall u B(z,u,c)$$
$$\mathsf{EA} \vdash \lim_{w\to\infty} f(c,w) = 1 \to \exists x \forall y \neg A(x,y,c)$$
$$I\Sigma^0_n \vdash \exists z \forall u B(z,u,c) \to \forall x \exists y A(x,y,c)$$

where the ordering $<$ denotes a standard ε_0-ordering $<_{\varepsilon_0}$.

First note that by Theorem 3 we have Σ^0_1-minimization up to each ordinal less than ω_{1+n} in $I\Sigma^0_n$. Hence $\exists \alpha <_{\varepsilon_0} K[\alpha = \min_{<_{\varepsilon_0}}\{\beta : \exists w[\beta = h(c,w)]\}]$. Pick a w so that the least $\alpha = h(c,w)$. Assuming that EA (a fortiori $I\Sigma^0_n$) proves (weakly descending) and (lowering), we have

$$I\Sigma^0_n \vdash \forall u \geq w[f(c,u) = f(c,w)].$$

Therefore the convergence of the predicate f is shown in $I\Sigma^0_n$. Also

$$I\Sigma^0_n \vdash \forall x \exists y A(x,y,c) \leftrightarrow \lim_{w\to\infty} f(c,w) = 0.$$

The converse follows from the following Reduction Theorem 5.

The theorem says that if a disjunction $\exists x \forall y \neg A(x,y,c) \vee \exists z \forall u B(z,u,c)$ of Σ^0_2-formulas is provable in $I\Sigma^0_n$, then one can construct an elementary recursive predicate f whose limit tells us which disjunct is true. The convergence of f is ensured by a descending function h in ordinals$< \omega_{1+n}$. Moreover these are all provable in EA.

Assuming the convergence of f(, which is provable in $I\Sigma^0_n$) this is a provable version of the classical Reduction Property of Σ^0_2 sets to Δ^0_2 sets. The point is that the Δ^0_2 sets $\{c : \lim_{w\to\infty} f(c,w) = 0\}$ are in a level $D_{<\omega_{1+n}}$ of Ershov hierarchy, demonstrably in $I\Sigma^0_n$.

Theorem 5 (Reduction Property) *Let $n \geq 1$.*

Suppose $I\Sigma^0_n \vdash \exists x \forall y \neg A(x,y,c) \vee \exists z \forall u B(z,u,c)$ for quantifier-free A, B. Then there exists an elementary *recursive predicate f, an ordinal $K < \omega_{1+n}$ and an* elementary *recursive function h such that*

1. (weakly descending)

$$\mathsf{EA} \vdash K > h(c,w) \geq h(c,w+1)$$

2. (lowering)

$$\mathsf{EA} \vdash f(c,w) \neq f(c,w+1) \to h(c,w) > h(c,w+1)$$

3. (reduction)

$$\mathsf{EA} \vdash \lim_{w\to\infty} f(c,w) = 0 \to \exists z\forall u B(z,u,c)$$
$$\mathsf{EA} \vdash \lim_{w\to\infty} f(c,w) = 1 \to \exists x\forall y \neg A(x,y,c)$$

In what follows, given a $I\Sigma^0_n$-proof of $\exists x\forall y\neg A(x,y,c) \lor \exists z\forall u B(z,u,c)$ let us construct a predicate f, an ordinal $K < \omega_{1+n}$ and a function h enjoying (weakly descending), (lowering) and (reduction).

Let $p(x,y,c)$ denote the characteristic function of the predicate

$$A((x)_0,(y)_0,c) \to B((x)_1,(y)_1,c),$$

where $(x)_i$ $(i = 0,1)$ denotes the projections of the pairing function.

Then

$$\exists x\forall y[p(x,y,c) = 0]$$

is provable in $I\Sigma^0_n$.

2.1. *Infinitary derivations*

In what follows let us consider (finite or infinite) derivations in one-sided sequent calculi. Given a finite derivation of $\exists x\forall y[p(x,y,c) = 0]$ in $I\Sigma^0_n$, first eliminate cut inferences partially to get a derivation of the same formula in which any cut formula is Σ^0_n.

Next embed the derivation into an infinite derivation of the sentence

$$\exists x\forall y[p(x,y,\bar{c}) = 0]$$

with the c-th numeral $\bar{c}$. Then eliminate cut inferences to get a cut-free derivation P_c of the same sentence. As usual the depth of P_c is bounded by an ordinal $K < \omega_{1+n}$ uniformly, i.e., $\forall c[\mathrm{dp}(P_c) < K]$.

In the derivation P_c, the *initial sequents* are

$$(Int)\ \Gamma, E$$

for true equation E. The equation E is called the *main formula* of the initial sequent.

In what follows we identify the closed term t with the numeral $\bar{n}$ of its value $n = val(t)$.

Note that the value of closed terms and truth values of equations in LEA are elementary recursively computable. The initial sequents are regarded as inference rules with empty premiss (upper sequent), and with the empty list of side formulas.

The *inference rules* are $(\exists), (\forall)$, and the repetition rule (Rep). These are standard ones.

$$\frac{\Gamma, B(\bar{n})}{\Gamma, \exists x B(x)}\,(\exists)\;;\quad \frac{\cdots\quad \Gamma, B(\bar{n})\quad \cdots (n \in \omega)}{\Gamma, \forall x B(x)}\,(\forall)\;;\quad \frac{\Gamma}{\Gamma}\,(Rep)$$

where $\exists x B(x)$ in the $(\exists)$ and $\forall x B(x)$ in the $(\forall)$ are the *main formula* of the inference, and $B(\bar{n})$ are *side formulas* of the inferences. The inference (Rep) has no main nor side formulas.

Our infinitary derivations are equipped with additional informations as in [6].

Definition 6 An *infinitary derivation* is a sextuple

$$D = (T, Seq, Rule, Mfml, Sfml, ord)$$

which enjoys the following conditions. The naked tree of D is denoted $T = T(D)$.

1. $T \subseteq {}^{<\omega}\omega$ is a tree with its root $\emptyset$ such that

$$a * \langle n \rangle \in T \,\&\, m < n \Rightarrow a * \langle m \rangle \in T.$$

2. $Seq(a)$ for $a \in T$ denotes the sequent situated at the node a. If $Seq(a)$ is a sequent Γ, then it is denoted

$$a : \Gamma.$$

3. $Rule(a)$ for $a \in T$ denotes the name of the inference rule with its lower sequent $Seq(a)$.

4. $Mfml(a)$ for $a \in T$ denotes the *main formula* of the inference rule $Rule(a)$. When $Rule(a) = (Rep)$, then $Mfml(a) = \emptyset$.

5. $Sfml(a*\langle n \rangle)$ for $a*\langle n \rangle \in T$ denotes the *side formula* of the inference rule $Rule(a)$, which is in the n-th upper sequent, i.e., $Sfml(a*\langle n \rangle) \in Seq(a * \langle n \rangle)$. When $Rule(a) = (Rep), (Int)$, then $Sfml(a * \langle n \rangle) = \emptyset$.

6. $ord(a)$ for $a \in T$ denotes the ordinal$<_{\varepsilon_0} K$ attached to a.

7. The sextuple $(T, Seq, Rule, Mfml, Sfml, ord)$ has to be locally correct with respect to inference rules of the infinitary calculus and for being well founded tree T.

In a derivation each inference rule except (Int) receives the following nodes:

$$\frac{a * \langle 0 \rangle : \Gamma, B(\bar{n})}{a : \Gamma, \exists x B(x)}\,(\exists)\;;\quad \frac{\cdots\quad a * \langle n \rangle : \Gamma, B(\bar{n})\quad \cdots (n \in \omega)}{a : \Gamma, \forall x B(x)}\,(\forall)\;;\quad \frac{a * \langle 0 \rangle : \Gamma}{a : \Gamma}\,(Rep)$$

The ordinals $ord_c(a)$ in the inference $(\forall)$

$$\frac{\cdots \quad a * \langle n \rangle : \Gamma, B(\bar{n}) \quad \cdots (n \in \omega)}{a : \Gamma, \forall x B(x)} \; (\forall)$$

enjoys

$$ord_c(a) > ord_c(a * \langle n \rangle) = ord_c(a * \langle m \rangle) \tag{5}$$

for any n, m.

As in [6] we see that the function $c \mapsto P_c$ is elementary recursive. We denote $P_c = (T_c, Seq_c, Rule_c, Mfml_c, Sfml_c, ord_c)$.

2.2. *Searching witnesses of Σ^0_2 in derivations*

Let us define a tracing function $\sigma(c, i) \in T_c = T(P_c)$.

The function $\{\sigma(c, w)\}_w$ indicates the trail in the proof tree T_c in which we go through in searching a witness x_a of $\exists x \forall y [p(x, y, \bar{c}) = 0]$, and verifying $\forall y [p(x_a, y, \bar{c}) = 0]$.

1. $\sigma(c, 0) = \emptyset(\text{root})$.
   In what follows let $a = \sigma(c, w)$.

2. Until $Seq_c(a)$ is an upper sequent of an $(\forall)$, go to the leftmost branch:

$$\sigma(c, w + 1) = a * \langle 0 \rangle.$$

   For example

$$\frac{a * \langle 0 \rangle : \Gamma, \exists x \forall y [p(x, y, \bar{c}) = 0], \forall y [p(x_a, y, \bar{c}) = 0]}{a : \Gamma, \exists x \forall y [p(x, y, \bar{c}) = 0]} \; (\exists)$$

3. The case when $Rule_c(b) = (\forall)$ with $a = b * \langle n \rangle$. Namely $Seq_c(a)$ is the n-th upper sequent of an $(\forall)$.

$$\frac{\cdots \quad a : \Gamma, p(x_a, y_a, \bar{c}) = 0 \quad \cdots}{\Gamma, \forall y [p(x_a, y, \bar{c}) = 0]} \; (\forall)$$

   x_a, y_a are closed terms.

   (a) If $p(x_a, y_a, \bar{c}) = 0$ is a TRUE equation, $\sigma(c, w + 1) = a \oplus 1$, the next right to the a:

$$\frac{\sigma(c, w) : \Gamma, p(x_a, y_a, \bar{c}) = 0 \quad \sigma(c, w + 1) : \Gamma, p(x_a, y_a + 1, \bar{c}) = 0}{\Gamma, \forall y [p(x_a, y, \bar{c}) = 0]} \; (\forall)$$

   where for an $a = (a_0, \ldots, a_{n-2}, a_{n-1}) \in {}^{<\omega}\omega$

$$a \oplus 1 = (a_0, \ldots, a_{n-2}, a_{n-1} + 1)$$

if $lh(a) = n > 0$.
$\emptyset \oplus 1$ is defined to be $\emptyset$.

(b) Otherwise $\sigma(c, w+1) = a * \langle 0 \rangle$, i.e., go to the leftmost branch from a.

$$\dfrac{\cdots \quad \dfrac{\sigma(c, w+1) : \Delta \quad \cdots}{\sigma(c, w) : \Gamma, p(x_a, y_a, \bar{c}) = 0} \quad \cdots}{\Gamma, \forall y[p(x_a, y, \bar{c}) = 0]} \ (\forall)$$

It is easy to see that the function $(c, w) \mapsto \sigma(c, w)$ is elementary recursive since $\max(\{(\sigma(c, w))_i : i < lh(\sigma(c, w))\} \cup \{lh(\sigma(c, w))\}) \leq w$.

Once $\sigma(c, w)$ is on an $(\forall)$, the tracing function goes through the upper sequents as long as the equations $p(x_a, y_a, \bar{c}) = 0$ is TRUE.

It is intuitively clear that after a finite number of steps, the sequence $\{\sigma(c, w)\}_w$ goes through the upper sequents of an $(\forall)$:

$$\dfrac{\sigma(c, w_0) : \Gamma, p(x_a, 0, \bar{c}) = 0 \quad \cdots \quad \sigma(c, w_0 + y) : \Gamma, p(x_a, \bar{y}, \bar{c}) = 0 \quad \cdots}{\Gamma, \forall y[p(x_a, y, \bar{c}) = 0]} \ (\forall)$$

since $\forall y[p(x_a, y, \bar{c}) = 0]$ is true for an x_a. We will know at the limit the fact, i.e., for $x = (x_a)_0$ and $z = (x_a)_1$

$$\exists y A(\bar{x}, y, \bar{c}) \to \forall u B(\bar{z}, u, \bar{c})$$

is true.

Now let us define an elementary recursive predicate f as follows.

1. $f(c, 0) = 1$.

2. Alternate values $f(c, w+1) = 1 - f(c, w)$ if $Seq_c(\sigma(c, w+1))$ is an upper sequent of an inference other than $(\forall)$.

3. Suppose $Seq_c(\sigma(c, w+1))$ is the n-th upper sequent of an $(\forall)$, and $\sigma(c, w+1) = b * \langle n \rangle$.

$$\dfrac{\cdots \quad b * \langle n \rangle : \Gamma, A(x_b, (\bar{n})_0, \bar{c}) \to B(z_b, (\bar{n})_1, \bar{c}) \quad \cdots}{b : \Gamma, \forall y, u[A(x_b, y, \bar{c}) \to B(z_b, u, \bar{c})]} \ (\forall)$$

$f(c, w+1) = 0$ iff $A(x_b, (n)_0, \bar{c}) \to B(z_b, (n)_1, \bar{c})$ is true, and the following condition holds:

$$\exists k \leq n[A(x_b, (\bar{k})_0, \bar{c})]$$

Namely

$$f(c, w+1) = 0 \Leftrightarrow$$
$$[A(x_b, (n)_0, \bar{c}) \to B(z_b, (n)_1, \bar{c})] \,\&\, \exists k \le n[A(x_b, (\bar{k})_0, \bar{c})]$$

Suppose $\sigma(c, w)$ is on an $(\forall)$. Until a witness k such that $A(x_b, (\bar{k})_0, \bar{c})$ is found, $f(c, w) = 1\,(w < k)$. After a witness k has been found, $f(c, w) = 0\,(w \ge k)$ as long as $A(x_b, (\bar{n})_0, \bar{c}) \to B(z_b, (\bar{n})_1, \bar{c})$ is true.

Therefore if the tracing function $\sigma(c, w)$ goes through the upper sequents of the $(\forall)$, then either $\lim_{w\to\infty} f(c, w) = 1$ and $\forall y \neg A(x_b, y, \bar{c})$, or $\lim_{w\to\infty} f(c, w) = 0$ and $\forall u B(z_b, u, \bar{c})$.

Proposition 7 *1. Suppose that $b * \langle n\rangle = \sigma(c, w+1)$ and $Seq_c(b * \langle n\rangle)$ is the n-th upper sequent of an inference $(\forall)$. Then $\{f(c, u) : \sigma(c, u) = b * \langle m\rangle, m \le n\}$ changes the values at most twice. Moreover if $f(c, u) = 0$ and $f(c, v) = 1$ for some $u < v \le w+1$, then $v = w+1$ and $\sigma(c, v+1) = \sigma(c, v) * \langle 0\rangle$, i.e., $Seq_c(\sigma(c, v))$ is the last upper sequent of the inference $(\forall)$ in the tracing function σ.*

2. (Reduction)

$$\mathsf{EA} \vdash \lim_{w\to\infty} f(c, w) = 0 \to \exists z \forall u B(z, u, c)$$
$$\mathsf{EA} \vdash \lim_{w\to\infty} f(c, w) = 1 \to \exists x \forall y \neg A(x, y, c)$$

Proof. Recall that an inference rule $(\forall)$ in P_c is of the form:

$$\frac{\ldots \quad b * \langle n\rangle : \Gamma, p(x_{b*\langle n\rangle}, \bar{n}, \bar{c}) = 0 \quad \ldots}{b : \Gamma, \forall y[p(x_b, y, \bar{c}) = 0]} \;(\forall)$$

where

$$p(x_{b*\langle n\rangle}, \bar{n}, \bar{c}) = 0 \leftrightarrow [A(((x)_{b*\langle n\rangle})_0, (\bar{n})_0, \bar{c}) \to B(((x)_{b*\langle k\rangle})_1, (\bar{n})_1, \bar{c})]$$

Let u be such that $\sigma(c, u) = b * \langle m\rangle$ with an $m \le n$. Then by the definition of the tracing function σ, we have for $m < n$ $p(x_{b*\langle m\rangle}, \bar{m}, \bar{c}) = 0$, i.e.,

$$A((x_{b*\langle m\rangle})_0, (\bar{m})_0, \bar{c}) \to B((x_{b*\langle m\rangle})_1, (\bar{m})_1, \bar{c}).$$

Suppose there exists a $u \le w+1$ such that $f(c, u) = 0$, and let u denote the minimal such one.

Then for any v with $u \le v < w+1$, we have $f(c, v) = 0$. Therefore if $f(c, v) = 1$ for a $v > u$, it must be the case $v = w+1$. This means that for some $k \leq n-1$ with $(k)_0 = (n)_0$

$$A((x_b)_0, (k)_0, c) \wedge \neg B((x_b)_1, (n)_1, c).$$

Hence $p(x_b, n, c) \neq 0$, and $\sigma(c, v+1) = \sigma(c, v) * \langle 0 \rangle$. □

Next define h as follows.

1. $$h(c, 0) = 3 \cdot ord_c(\emptyset).$$

   In what follows put $a = \sigma(c, w+1)$ and let $Seq_c(a)$ be an upper sequent of an inference $Rule_c(b)$ with $a = b * \langle n \rangle$.

2. The case when $Rule_c(b)$ is an inference rule other than $(\forall)$.

   $$h(c, w+1) := 3 \cdot ord_c(\sigma(c, w+1)).$$

   By Proposition 7.1 we know that the $f(c, u)$ changes the values at most twice in the upper sequents of an $(\forall)$.

3. The case when $n = 0$ and $Rule_c(b) = (\forall)$.

   $$h(c, w+1) := 3 \cdot ord_c(\sigma(c, w+1)) + 2.$$

4. The case when $n > 0$, $Rule_c(b) = (\forall)$.

   $$\frac{\cdots \quad b * \langle n \rangle : \Gamma, A(x_b, (\bar{n})_0, \bar{c}) \to B(z_b, (\bar{n})_1, \bar{c}) \quad \cdots}{b : \Gamma, \forall y, u[A(x_b, y, \bar{c}) \to B(z_b, u, \bar{c})]} \ (\forall)$$

   We have by (5)

   $$ord_c(\sigma(c, w)) = ord_c(\sigma(c, w+1)).$$

   (a) The case when $f(c, w+1) = f(c, w)$.

   $$h(c, w+1) := h(c, w).$$

   where $\sigma(c, w+1) = \sigma(c, w) \oplus 1$.

   (b) The case when $f(c, w) = 1 \,\&\, f(c, w+1) = 0$.
   Then $\sigma(c, w+1) = \sigma(c, w) \oplus 1$ and $n = \min\{k : A(x_b, (\bar{k})_0, \bar{c})\}$.
   Let

   $$h(c, w+1) := 3 \cdot ord_c(\sigma(c, w+1)) + 1.$$

   (c) The case when $f(c, w) = 0 \,\&\, f(c, w+1) = 1$.
   This means that $A(x_b, (\bar{n})_0, \bar{c}) \to B(z_b, (\bar{n})_1, \bar{c})$ is FALSE and $\sigma(c, w+2) = \sigma(c, w+1) * \langle 0 \rangle$.

$$h(c, w+1) := 3 \cdot ord_c(\sigma(c, w+1)).$$

Obviously h is elementary recursive.

Proposition 8

$$\text{(weakly descending) } \mathsf{EA} \vdash 3K > h(c, w) \geq h(c, w+1)$$

$$\text{(lowering) } \mathsf{EA} \vdash f(c, w) \neq f(c, w+1) \rightarrow h(c, w) > h(c, w+1)$$

Proof. (weakly descending) is obvious.

Consider the case when $\sigma(c, w+1) = a$ and $Seq_c(a)$ is an upper sequent of an inference $Rule_c(b) = (\forall)$ with $a = b * \langle n \rangle$.

If $n = 0$, then

$$h(c, w+1) = 3 \cdot ord_c(\sigma(c, w+1)) + 2 < 3 \cdot ord_c(\sigma(c, w)) \leq h(c, w)$$

since $Seq_c(\sigma(c, w))$ is the lower sequent of $Seq_c(a)$ with $b = \sigma(c, w)$.

Assume $n > 0$. Using Proposition 7.1 we see $h(c, w+1) \in \{3 \cdot ord_c(\sigma(c, w+1)) + i : i < 3\}$.

Moreover if $\sigma(c, w+2) = \sigma(c, w+1) * \langle 0 \rangle$, then

$$h(c, w+1) \geq 3 \cdot ord_c(\sigma(c, w+1)) > 3 \cdot ord_c(\sigma(c, w+1)) + 2 \geq h(c, w+2).$$

□

This completes a proof of Theorems 5 and 4.

2.3. *Provably Σ_2^0-functions*

If $\exists z \forall u B(z, u, c)$ is provable for quantifier-free B, then we can find a witness $z = \lim_{w \to \infty} f(c, w)$ as limits of an f, whose convergence is ensured by weakly descending chains of ordinals.

Theorem 9 *Suppose $I\Sigma_n^0 \vdash \exists z \forall u B(z, u, c)$ for quantifier-free B. Then there exist* elementary *recursive functions f, h and an ordinal $K < \omega_{1+n}$ such that*

1.

$$\text{(weakly descending) } \mathsf{EA} \vdash K > h(c, w) \geq h(c, w+1)$$

2.

$$\text{(lowering) } \mathsf{EA} \vdash f(c, w) \neq f(c, w+1) \rightarrow h(c, w) > h(c, w+1)$$

3.

$$\mathsf{EA} \vdash \lim_{w \to \infty} f(c, w) = z \rightarrow \forall u B(z, u, c)$$

Proof. As in the proof of Theorem 5, let us define a tracing function σ.

$\sigma(c, w)$ goes on the leftmost branch up to an $(\forall)$. $\sigma(c, w)$ goes through the upper sequents of $(\forall)$ as long as side formulas $B(z_a, \bar{n}, \bar{c})$ is TRUE. If a FALSE side formula $B(z_a, \bar{n}, \bar{c})$ is found, then throw z_a away and go on the leftmost branch.

Now h is defined by $h(c, w) := ord_c(\sigma(c, w))$. f is defined obviously. $f(c, w) = z_a$ if $Seq_c(\sigma(c, w))$ is an upper sequent of an $(\forall)$ with its side formula $B(z_a, \bar{n}, \bar{c})$. Otherwise $f(c, w)$ is arbitrary, say $f(c, w) = 0$. □

It is well known that the 1-consistency $\mathrm{RFN}_{\Pi^0_2}(I\Sigma^0_n)$ is equivalent over Primitive Recursive Arithmetic PRA to the fact that there is no primitive recursive descending chain of ordinals$< \omega_{1+n}$.

Theorem 10 *(Cf. [1] for another form of the 2-consistency of arithmetic.)*

The 2-consistency $\mathrm{RFN}_{\Pi^0_3}(I\Sigma^0_n)$ *is equivalent over PRA to the fact that every primitive recursive weakly descending chain of ordinals*$< \omega_{1+n}$ *has a limit, or equivalently to the fact that for any primitive recursive sequence* $\{h(c, w)\}_w$ *of ordinals*$< \omega_{1+n}$ *the least ordinal* $\min_{<_{\varepsilon_0}}\{h(c, w) : w \in \omega\}$ *exists.*

Proof.

Over PRA, $\mathrm{RFN}_{\Pi^0_3}(I\Sigma^0_n)$ yields the existence of the least ordinal $\min_{<_{\varepsilon_0}}\{h(c, w) < \omega_{1+n} : w \in \omega\}$ since $\alpha = \min_{<_{\varepsilon_0}}\{\beta : \exists w[\beta = h(c, w)]\}$ is a Σ^0_2-formula.

Conversely let $f(c, w) < 2$ be defined as follows:

1. c is not a Gödel number of an $I\Sigma^0_n$-proof of a Σ^0_2-sentence: Then $f(c, w) = 0$ for any w.
2. c is a Gödel number of an $I\Sigma^0_n$-proof of a Σ^0_2-sentence $\exists z \forall u B_c(z, u)$: $f(c, w)$ is defined as in Theorem 9 for a cut free infinite derivation P_c of $\exists z \forall u B_c(z, u)$. Note that f is non-elementary since it involves cut elimination for predicate logic.

Also let $h(c, w) := ord_c(\sigma(c, w))$.

Then

1.

$$\text{(weakly descending) PRA} \vdash \omega_{1+n} > h(c, w) \geq h(c, w + 1)$$

2.

$$\text{(lowering) PRA} \vdash f(c, w) \neq f(c, w + 1) \rightarrow h(c, w) > h(c, w + 1)$$

3.

$$\mathsf{PRA} \vdash \lim_{w\to\infty} f(c,w) = 0 \to \mathrm{Prov}_{I\Sigma^0_n}(c, \ulcorner \exists z \forall u B_c(z,u) \urcorner) \to \exists z \forall u B_c(z,u)$$

Therefore

$$\mathsf{PRA} \vdash \forall c[\exists w \forall u \geq w\{h(c,u) = h(c,w)\} \to \exists \ell\{\lim_{w\to\infty} f(c,w) = \ell\}]$$

and

$$\mathsf{PRA} \vdash \forall c \exists \ell[\lim_{w\to\infty} f(c,w) = \ell] \to \mathrm{RFN}_{\Pi^0_3}(I\Sigma^0_n).$$

□

3. Provably Δ^0_2 in EA

In this section we consider the Δ^0_2-sets provably in EA.

The following Theorem 11 states that for a set is provably Δ^0_2 in EA iff it is EA-provably in the class D_n of a finite level in the Ershov hierarchy. The finite levels $\{D_n : n < \omega\}$ are called the difference (or Boolean) hierarchy, and by a result due to H. Putnam(Theorem 2 in [7]) we see that a set is provably Δ^0_2 in EA iff it is equivalent to a Boolean combination of Σ^0_1-formulas, provably in EA. This answers to a problem of L. Beklemishev.

Theorem 11 *The following are equivalent for quantifier-free A, B and a free variable c.*

1. EA *proves*

$$\forall x \exists y A(x,y,c) \leftrightarrow \exists z \forall u B(z,u,c) \tag{4}$$

2. *There exists a binary* elementary *recursive predicate f, a natural number $K < \omega$ and an* elementary *recursive function $h : \omega \times \omega \to K$ such that*

   (a) (weakly descending)

$$\mathsf{EA} \vdash K > h(c,w) \geq h(c,w+1)$$

   (b) (lowering)

$$\mathsf{EA} \vdash f(c,w) \neq f(c,w+1) \to h(c,w) > h(c,w+1)$$

   (c) (reduction)

$$\mathsf{EA} \vdash \lim_{w\to\infty} f(c,w) = 0 \to \exists z \forall u B(z,u,c)$$
$$\mathsf{EA} \vdash \lim_{w\to\infty} f(c,w) = 1 \to \exists x \forall y \neg A(x,y,c)$$
$$\mathsf{EA} \vdash \exists z \forall u B(z,u,c) \to \forall x \exists y A(x,y,c)$$

   for the usual ordering $<$ on ω.

Proof. Assume EA proves (weakly descending) and (lowering) for a natural number K. Then EA also proves the convergence of f:

$$\mathsf{EA} \vdash \exists \ell[\lim_{w\to\infty} f(c,w) = \ell],$$

(reduction) yields

$$\mathsf{EA} \vdash \forall x \exists y A(x,y,c) \leftrightarrow \lim_{w\to\infty} f(c,w) = 0.$$

Conversely suppose that EA proves (4). Then so is the $\exists\forall$-formula

$$\exists x \exists z \forall y \forall z [A(x,y,c) \to B(z,u,c)].$$

By the Herbrand's theorem there exist a list of variables $\{a_i, b_i : i \leq r\}$ and a list of terms $\{t_i, s_i : i \leq r\}$ such that

$$\bigvee\{A(t_i,a_i,c) \to B(s_i,b_i,c) : i \leq r\} \tag{6}$$

is provable in EA, and variables occurring in t_i, s_i are among a_j, b_j for $j < i$ besides the parameter c.

For simplicity consider the case when $r = 1$. Then we have

$$\mathsf{EA} \vdash \neg A(t_0,a_0,c) \vee B(s_0,b_0,c) \vee \neg A(t_1(a_0,b_0),a_1,c) \vee B(s_1(a_0,b_0),b_1,c) \tag{7}$$

Let f denote the elementary recursive predicate

$$f(c,w) := \begin{cases} 0 \ [\{t_0 \leq w \wedge \exists y \leq w A(t_0,y,c)\} \wedge \{s_0 \leq w \wedge \forall u \leq w B(s_0,u,c)\}] \vee \\ \quad [\exists a_0, b_0 \leq w\{A(t_0,a_0,c) \wedge \neg B(s_0,b_0,c) \wedge \\ \quad \exists a_1 \leq w A(t_1(a_0,b_0),a_1,c) \wedge \forall b_1 \leq w B(s_1(a_0,b_0),b_1,c)\}] \\ 1 \ \text{otherwise} \end{cases}$$

For the number

$$K := 1 + 2r + 2 (= 5 \text{ if } r = 1),$$

let $h : \omega \times \omega \to K$ denote the elementary recursive function

$$h(c,0) = K - 1$$

and

$$h(c,w+1) := \begin{cases} h(c,w) & \text{if } f(c,w+1) = f(c,w) \\ h(c,w) \dot{-} 1 & \text{if } f(c,w+1) \neq f(c,w) \end{cases}$$

Lemma 12 EA *proves the facts (weakly descending), (lowering) and (reduction).*

Proof. Argue in EA. (weakly descending) is obvious.

Suppose

$$\lim_{w\to\infty} f(c,w) = \ell$$

for an $\ell = 0, 1$.

By (4) we have either $\exists z \forall u B(z,u,c)$ or $\exists x \forall y \neg A(x,y,c)$.

First consider the case when $\exists z \forall u B(z,u,c)$. Then $\forall x \exists y A(x,y,c)$. Hence by (7) either $\forall b_0 B(s_0, b_0, c)$ or $\forall b_1 B(s_1(a_0,b_0), b_1, c)$ for some a_0, b_0 with $A(t_0, a_0, c) \wedge \neg B(s_0, b_0, c)$.

If $\forall b_0 B(s_0, b_0, c)$, then $f(c,w) = 0$ for any $w \geq \max\{t_0, s_0, y_0\}$, where $y_0 = \mu y.A(t_0, y.c)$. Therefore $\ell = 0$. Moreover $f(c,w) = 1$ for $w < \max\{t_0, s_0, y_0\}$.

Next assume $\forall b_1 B(s_1(a_0,b_0), b_1, c)$ for the minimal $a_0.b_0$ such that $A(t_0, a_0, c) \wedge \neg B(s_0, b_0, c)$. Then let a_1 denote the minimal a_1 such that $A(t_1(a_0,b_0), a_1, c)$. We have $f(c,w) = 0$ for any $w \geq \max\{a_0, b_0, a_1\}$, and hence $\ell = 0$.

Now consider $w < \max\{a_0, b_0, a_1\}$. Then $f(c,w) = 0$ iff $\max\{t_0, s_0, a_0\} \leq w < b_0$. Therefore $\lambda w.f(c,w)$ changes its values at most three times (when $\max\{t_0, s_0, a_0\} < b_0 < a_1$).

Next consider the case when $\exists x \forall y \neg A(x,y,c)$. We have $\forall z \exists u \neg B(z,u,c)$. Then $f(c,w) = 1$ for any w if $\forall a_0 \neg A(t_0, a_0, c)$, and $f(c,w) = 1$ for any $w \geq \max\{b_0, b_1\}$ if $\forall a_1 \neg A(t_1(a_0,b_0), a_1, c)$ for the minimal a_0, b_0, b_1 such that $A(t_0, a_0, c) \wedge \neg B(s_0, b_0, c)$ and $\neg B(s_1(a_0,b_0), b_1, c)$. Therefore $\ell = 1$.

Finally assume $\forall a_1 \neg A(t_1(a_0,b_0), a_1, c)$, and consider $w < \max\{b_0, b_1\}$. Then $f(c,w) = 0$ iff $\max\{t_0, s_0, a_0\} \leq w < b_0$. Therefore $\lambda w.f(c,w)$ changes its values at most two times in this case.

In any cases, (reduction) was shown, and $\lambda w.f(c,w)$ changes its values at most $1 + 2r (= 3$ if $r = 1)$ times for any c, i.e.,

$$\forall (w_0 < w_1 < \cdots < w_{1+2r}) \exists i \leq 1 + 2r [f(c, w_i) = f(c, w_i + 1)].$$

Hence (lowering) follows. □

Lemma 12 with a result due to H. Putnam(Theorem 2 in [7]) yields the

Theorem 13 *Suppose that* EA *proves*

$$\forall x \exists y A(x,y,c) \leftrightarrow \exists z \forall u B(z,u,c)$$

for quantifier-free A, B.

Then over EA $\exists z \forall u B(z,u,c)$ *is equivalent to a Boolean combination of* Σ^0_1*-formulas.*

Proof. (cf. [7].) Let r be as in (6). Define Σ^0_1 $Y_k(c)$ $(k \leq 1 + 2r)$ and $N_i(c)$ by

$$\begin{aligned} Y_k(c) &:\Leftrightarrow \exists(w_0 < w_1 < \cdots < w_{k-1})\forall i < k[f(c, w_i) \neq f(c, w_i + 1) \\ &\qquad \wedge f(c, w_{k-1}) = 0] \\ N_k(c) &:\Leftrightarrow \exists(w_0 < w_1 < \cdots < w_{k-1})\forall i < k[f(c, w_i) \neq f(c, w_i + 1) \\ &\qquad \wedge f(c, w_{k-1}) = 1]] \end{aligned}$$

for $k > 0$, and $Y_0(c) :\Leftrightarrow f(c, 0) = 0$, $N_0(c) :\Leftrightarrow f(c, 0) = 1$. Also put $N_{2r+2}(c) :\Leftrightarrow 0 = 1$.

Then EA proves that

$$\exists z \forall u B(z, u, c) \leftrightarrow \bigvee\{Y_k(c) \wedge \neg N_{k+1}(c) : k \leq 1 + 2r\}.$$

□

As in Theorems 9, 10 we see the following theorems.

Theorem 14 *Suppose* $\mathsf{EA} \vdash \exists z \forall u B(z, u, c)$ *for quantifier-free* B. *Then there exist* elementary *recursive functions* f, h *and a natural number* $K < \omega$ *such that*

1.

$$\text{(weakly descending) } \mathsf{EA} \vdash K > h(c, w) \geq h(c, w + 1)$$

2.

$$\text{(lowering) } \mathsf{EA} \vdash f(c, w) \neq f(c, w + 1) \rightarrow h(c, w) > h(c, w + 1)$$

3.

$$\mathsf{EA} \vdash \lim_{w \to \infty} f(c, w) = z \rightarrow \forall u B(z, u, c)$$

Theorem 15 *The 2-consistency* $\mathrm{RFN}_{\Pi^0_3}(\mathsf{EA})$ *is equivalent over* PRA *to the fact that every primitive recursive weakly descending chain of natural number*$< \omega$ *has a limit, or equivalently to the fact that for any primitive recursive sequence* $\{h(c, w)\}_w$ *of natural number*$< \omega$ *the least number* $\min_<\{h(c, w) < \omega : w \in \omega\}$ *exists.*

Remark.

Obviously Theorems 11, 13 and 14 hold for any purely universal extension of EA, eg., $\mathsf{EA}+\mathrm{CON}(\mathsf{EA})$, PRA.

A. Nested limit existence rules

Every fragment in the Appendix is an extension of Elementary Recursive Arithmetic EA.

In [2], Beklemishev and Visser gave an elegant axiomatization of Σ^0_2-consequences of $I\Sigma^0_1$ in terms of the inference rule $(LimR)$ for limit existence principle:

$$\frac{\exists m \forall n \geq m\, h(n+1) \leq h(n)}{\exists m \forall n \geq m\, h(n) = h(m)}\ (LimR)$$

Moreover unnested applications of $(LimR)$ is shown to be equivalent to $I\Pi^-_1$ (over EA).

This reminds us another axiomatization of Σ^0_2-consequences of $I\Sigma^0_1$ in [5]. Namely $I\Sigma^0_1$ is a Σ^0_2 conservative extension of $L\Sigma^{-(\infty)}_1 = \bigcup_k L\Sigma^{-(k)}_1$, where $L\Sigma^{-(k)}_1$ denotes the schema

$$\exists x_1 \cdots \exists x_k \theta(x_1, \ldots, x_k) \to$$
$$\exists x_1 \cdots \exists x_k \bigwedge_{i=1}^{k} [\exists \vec{y} \theta(x_1, \ldots, x_i, \vec{y}) \wedge \forall z < x_i \forall \vec{y} \neg \theta(x_1, \ldots, x_{i-1}, z, \vec{y})]$$

for $\theta \in \Sigma^0_1$ without parameters.

For example $L\Sigma^{-(0)}_1 = \mathsf{EA}$ and $L\Sigma^{-(1)}_1 = L\Sigma^-_1 = I\Pi^-_1$.

In this Appendix we show that $L\Sigma^{-(k)}_1$ is equivalent to the k-nested applications of $(LimR)$. To be precise, let $(LimR)^{(k)} \vdash$ denote the derivability in the k-nested applications of $(LimR)$: $(LimR)^{(0)} \vdash$ is nothing but $\mathsf{EA} \vdash$, and if $(LimR)^{(k)} \vdash \exists m \forall n \geq m\, h(n+1) \leq h(n)$, then $(LimR)^{(k+1)} \vdash \exists m \forall n \geq m\, h(n) = h(m)$.

Theorem 16 *$(LimR)^{(k)} \vdash \varphi \Leftrightarrow L\Sigma^{-(k)}_1 \vdash \varphi$ for any φ.*

This is shown by induction on k. The proof is obtained by a slight modification of proofs in [2].

First consider

$$(LimR)^{(k)} \vdash L\Sigma^{-(k)}_1.$$

Let $<^{(k)}$ $(k \geq 1)$ denote the lexicographic order on k-tuples of natural numbers. Also $\langle x_1, \ldots, x_k \rangle^{(k)}$ denotes a(n elementary recursive) bijective coding of k-tuples with its inverses $(n)^{(k)}_i$ $(1 \leq i \leq k)$. In what follows the super scripts (k) are omitted.

Then $L\Sigma^{-(k)}_1$ says that if there exists an x satisfying $\varphi(x) \equiv \theta((x)_1, \ldots, (x)_k)$, then there exists a minimal such x with respect to $<^{(k)}$.

We can assume that EA proves

$$\exists i[\forall j \neq i(x_j = y_j) \wedge x_i < y_i] \to \langle x_1, \ldots, x_k \rangle < \langle y_1, \ldots, y_k \rangle \tag{8}$$

Now given a Δ^0_0-formula $\varphi(x_1, \ldots, x_k, x_{k+1})$ without parameters, we want to show $L\Sigma_1^{-(k)}$ with $\theta \equiv \exists x_{k+1}\varphi$.

As in [2] some elementary functions g_1, g, h, h' are defined successively as follows.

$$g_1(n) = \begin{cases} n & \text{if } \forall y \leq n \neg\varphi((y)_1, \ldots, (y)_k, (y)_{k+1}) \\ \langle (y)_1, \ldots, (y)_k \rangle & \text{otherwise with } y = \mu y \leq n\varphi((y)_1, \ldots, (y)_k, (y)_{k+1}) \end{cases}$$

$$g(n) = \begin{cases} \langle (n)_1, \ldots, (n)_k \rangle & \text{if } \exists u \leq (n)_{k+1}\varphi((n)_1, \ldots, (n)_k, u) \\ g_1(n) & \text{otherwise} \end{cases}$$

$h(0) = g(0)$ and

$$h(n+1) = \begin{cases} g(n+1) & \text{if } \forall k, m \leq n(k \neq m \to g(k) \neq g(m)) \\ g(n+1) & \text{if } \exists m \leq n(g(n+1) = g(m)) \text{ and } g(n+1) <^{(k)} h(n) \\ h(n) & \text{otherwise} \end{cases}$$

Observe that $h(n) \leq \max\{g(m) : m \leq n\}$, and hence h is elementary.

$$h'(x) = \begin{cases} h(x) & \text{if } \exists n \leq x\varphi((n)_1, \ldots, (n)_k, (n)_{k+1}) \\ 0 & \text{otherwise} \end{cases}$$

Then EA proves that h' is eventually decreasing with respect to $<^{(k)}$: $\exists m \forall n \geq m(h'(n+1) \leq^{(k)} h'(n))$. Therefore $h'_1(n) = (h'(n))_1$ is eventually decreasing. Hence $\exists y_1[y_1 = \lim_{x\to\infty} h'_1(x)]$ in $(LimR)^{(1)}$.

This in turn implies that $\langle (h'(n))_2, \ldots, (h'(n))_k \rangle$ is eventually decreasing with respect to $<^{(k-1)}$. Therefore $h'_2(n) = (h'(n))_2$ is eventually decreasing demonstrably in $(LimR)^{(1)}$. Hence $\exists y_2[y_2 = \lim_{x\to\infty} h'_2(x)]$ in $(LimR)^{(2)}$, and so on. Therefore $\exists y[y = \lim_{x\to\infty} h'(x)]$ in $(LimR)^{(k)}$.

Now assuming $\exists x_1 \cdots \exists x_k \exists x_{k+1}\theta(x_1, \ldots, x_k, x_{k+1})$, we see as in [2] that $y = \lim_{x\to\infty} h'(x) = \lim_{x\to\infty} h(x)$, and the limit y is the minimum of $\{\langle x_1, \ldots, x_k \rangle : \exists x_{k+1}\theta(x_1, \ldots, x_k, x_{k+1})\}$ with respect to the lexicographic order $<^{(k)}$ as desired.

Next assume by IH that

$$L\Sigma_1^{-(k)} \vdash \exists m \forall n \geq m\, h(n+1) \leq h(n).$$

We need to show

$$L\Sigma_1^{-(k+1)} \vdash \exists m \forall n \geq m\, h(n) = h(m).$$

For simplicity consider the case $k = 1$, and assume that EA proves that

$$\{\exists x_1 \varphi_1(x_1) \to \exists x_1[\varphi_1(x_1) \land \forall z < x_1 \neg\varphi_1(z)]\} \land$$
$$\{\exists x_2 \varphi_2(x_2) \to \exists x_2[\varphi_2(x_2) \land \forall z < x_2 \neg\varphi_2(z)]\}$$
$$\to \exists m \forall n \geq m\, h(n+1) \leq h(n)$$

Let $\varphi_i(x_i) \equiv \exists y \theta_i(x_i, y)$.

By the Herbrand's Theorem there exists a sequence of terms $m_0(a_1, a_2, b_1, b_2)$, $m_1(x_0, a_1, a_2, b_1, b_2)$, $m_2(x_0, x_1, a_1, a_2, b_1, b_2), \ldots, m_k(x_0, \ldots, x_{k-1}, a_1, a_2, b_1, b_2)$ such that the following disjunction is provable in EA:

$$\begin{aligned}
&\{\exists x_1 \varphi_1(x_1) \land [\neg\theta_1(a_1, b_1) \lor \exists z < a_1 \varphi_1(z)]\} \lor \\
&\{\exists x_2 \varphi_2(x_2) \land [\neg\theta_2(a_2, b_2) \lor \exists z < a_2 \varphi_2(z)]\} \lor \\
&(x_0 \geq m_0(a_1, a_2, b_1, b_2) \to h(x_0) \geq h(x_0+1)) \lor \\
&(x_1 \geq m_1(x_0, a_1, a_2, b_1, b_2) \to h(x_1) \geq h(x_1+1)) \lor \\
&\ldots \\
&(x_k \geq m_k(x_0, \ldots, x_{k-1}, a_1, a_2, b_1, b_2) \to h(x_k) \geq h(x_k+1))
\end{aligned} \tag{9}$$

First assume $\exists x_1 \varphi_1(x_1) \land \exists x_2 \varphi_2(x_2)$, and pick a minimal $a = \langle a_1, a_2 \rangle$ such that $\varphi_1(a_1) \land \varphi_2(a_2)$ and $\forall b < a \neg[\varphi_1((b)_1) \land \varphi_2((b)_2)]$. Then from (8) we see that $\forall z < a_i \neg\varphi_i(z)$ for $i = 1, 2$.

Now let

$$y_0 = \mu y_0[\exists b_1, b_2 \exists x_0 \geq m_0(a_1, a_2, b_1, b_2)\{\theta_1(a_1, b_1) \land \theta_2(a_2, b_2) \land h(x_0) = y_0\}]$$

by $L\Sigma_1^{-(2)}$.

If $\forall x_0 \geq m_0(a_1, a_2, b_1, b_2)(h(x_0) \geq h(x_0+1))$ for some b_1, b_2 with $\theta_1(a_1, b_1) \land \theta_2(a_2, b_2)$, then $y_0 = \lim_{x\to\infty} h(x)$.

Otherwise let

$$\begin{aligned}
y_1 = \mu y_1[&\exists b_1, b_2 \exists x_0 \geq m_0(a_1, a_2, b_1, b_2) \exists x_1 \geq m_1(x_0, a_1, a_2, b_1, b_2) \\
&(\theta_1(a_1, b_1) \land \theta_2(a_2, b_2) \land h(x_0) < h(x_0+1) \land h(x_1) = y_1)]
\end{aligned}$$

If $\forall x_1 \geq m_1(x_0, a_1, a_2, b_1, b_2)(h(x_1) \geq h(x_1+1))$ for some b_1, b_2 with $\theta_1(a_1, b_1) \land \theta_2(a_2, b_2)$, and an $x_0 \geq m_0(a_1, a_2, b_1, b_2)$, then $y_1 = \lim_{x\to\infty} h(x)$, and so on.

If $\neg\exists x_1 \varphi_1(x_1)$, then substitute 0 for a_1, b_1 in (9).

References

[1] T. Arai and G. Mints, Extended normal form theorems for logical proofs from axioms, Theor. Comp. Sci. 232 (2000), 121-132.

[2] L. D. Beklemishev and A. Visser, On the limit existence principles in elementary arithmetic and Σ_n^0-consequences of theories, Ann. Pure Appl. Logic 136(2005), 56-74.

[3] Y. l. Ershov, A certain hierarchy of sets. II. Algebra i Logika 7(1968), 15-47.

[4] P. Hájek and P. Pudlák, Metamathematics of First Order Arithmetic, Springer, 1993.

[5] R. Kaye, J. Paris and C. Dimitracopoulos, On parameter free induction schemas, Jour. Symb. Logic 53(1988), 1082-1097.

[6] G.E. Mints, Finite investigations of transfinite derivations, in: Selected Papers in Proof Theory (Bibliopolis, Napoli, 1992) 17-72.

[7] H. Putnam, Trial and error predicates and the solution to a problem of Mostowski, Jour. Symb. Logic 30(1965), 51-57.

[8] H. E. Rose, Subrecursion:Functions and hierarchies. Oxford Logic Guides 9. Oxford University Press 1984.

[9] Th. Skolem, Proof of some theorems on recursively enumerable sets. Notre Dame J. of Formal Logic 3, 65-74 (1962).

[10] R. Sommer, Transfinite inducton within Peano arithmetic, Ann. Pure Appl. Logic 76 (1995), 231-289.

[11] F. Stephan, Y. Yang and L. Yu, Turing degrees and the Ershov hierarchy, In: T. Arai, J. Brendle, C. T. Chong, R. Downey, Q. Feng, H. Kikyo and H. Ono (eds.), Proceedings of the 10th Asian Logic Conference, 2009, World Scientific, Singapore, pp. 300-321.

AMALGAMATION, ABSOLUTENESS, AND CATEGORICITY

John T. Baldwin*

University of Illinois at Chicago

We describe the major result on categoricity in $L_{\omega_1,\omega}$, placing it in the context of more general work in abstract elementary classes. In particular, we illustrate the role of higher dimensional amalgamations and sketch the role of a weak extension of ZFC in the proof. We expound the translation of the problem to studying atomic models of first order theories. We provide a simple example of the failure of amalgamation for a complete sentence of $L_{\omega_1,\omega}$. We prove some basic results on the absoluteness of various concepts in the model theory of $L_{\omega_1,\omega}$ and publicize the problem of absoluteness of $\aleph_1$-categoricity in this context. Stemming from this analysis, we prove Theorem: The class of countable models whose automorphism groups admit a complete left invariant metric is Π^1_1 but not Σ^1_1.

The study of infinitary logic dates from the 1920's. Our focus here is primarily on the work of Shelah using stability theoretic methods in the field (beginning with [She75]). In the first four sections we place this work in the much broader context of abstract elementary classes (aec), but do not develop that subject here. The main result discussed, Shelah's categoricity transfer theorem for $L_{\omega_1,\omega}$, explicitly uses a weak form of the GCH. This raises questions about the absoluteness of fundamental notions in infinitary model theory. Sections 5-7 and the appendix due to David Marker describe the complexity and thus the absoluteness of such basic notions as satisfiability, completeness, ω-stability, and excellence.[1] We state the question, framed in this incisive way by Laskowski, of the absoluteness of $\aleph_1$-categoricity. And from the model theoretic characterization of non-extendible models we derive the theorem stated in the abstract on the complexity of automorphism groups. Most of the results reported here in Sections 1-4 are due to Shelah; the many references to [Bal09] are to provide access to a unified exposition. I

*This article is a synthesis of the paper given in Singapore with later talks, including the Mittag-Leffler Institute in 2009 and CRM Barcelona in 2010. It reflects discussions with set theorists during my stay at Mittag-Leffler and discussion with the Infinity project members in Barcelona. The author wishes to thank the John Templeton Foundation for its support through Project #13152, Myriad Aspects of Infinity, hosted during 2009-2011 at the Centre de Recerca Matematica, Bellaterra, Spain. Baldwin was partially supported by NSF-0500841.

[1]David Marker is partially supported by National Science Foundation grant DMS-0653484.

don't know anywhere that the results in Section 5 have been published. The techniques are standard. Our main goal was to provide a reference for this material; but the distinction between the complexity of various notions for atomic classes as opposed to sentences of $L_{\omega_1,\omega}$ seems to be a new observation. The result in Section 6 is new but easy.

1. The Universe is Wide or Deep

Shelah made the following rough conjecture: Let $\boldsymbol{K}$ be a *reasonable* class of models.

Either for some λ, there are many models of cardinality λ or there are models of arbitrarily large cardinality.

Our metaphor requires some explanation. 'The universe' should perhaps be 'each universe'; universe refers to all models in a specific class. Further we are taking 'or' in the inclusive sense. Certainly, there are classes (e.g. dense linear orders) which are both wide and deep. Perhaps, taking narrow, as meaning there are few models in each cardinality, the aphorism better reads. A narrow universe is deep. It turns out that this question depends very much on the choice of 'reasonable'. It also seems to be sensitive to the choice of axioms of set theory. In order to give a precise formulation of the conjecture we have to specify 'many' and the notion of a 'reasonable class'. In general 'many' should mean 2^λ; but in important cases that have been proved, it is slightly smaller.

As is often the case there are some simplifying assumptions in this area that have been internalized by specialists but obscure the issues for other logicians. We try to explain a few of these simplifications and sketch some of the major results.

Some historical background will help clarify the issues. Much model theoretic research in the 60's focussed on general properties of first order and infinitary logic. A number of results seemed to depend heavily on extensions of ZFC. For example, both Keisler's proof that two structures are elementarily equivalent if and only if they have isomorphic ultrapowers and Chang's proof of two cardinal transfer required GCH. In general, even the existence of saturated models depends on the GCH. Shelah removed the set-theoretic hypothesis from Keisler's theorem. But various versions of two cardinal transfer were proven to require GCH and even large cardinal hypotheses. See [CK73].

The invention of stability theory radically recast the subject of model theory. E.g., for various classes in the stability hierarchy, it is straightforward to characterize in ZFC exactly in which cardinals there are saturated models. And for the best behaved theories the answers is: all cardinals. Further, for countable stable theories Shelah and Lachlan independently showed that two cardinal transfer between any pair of cardinalities is true in ZFC. Moreover, the fundamental notions of first order stability theory are absolute.

For first order logic, our guiding question is trivial[2]. If a theory has an infinite model then it has arbitrarily large models. The question is interesting for theories in logics which fail the upward Löwenheim-Skolem theorem. The notion of an Abstract Elementary Class (AEC) provides a general framework for analyzing such classes. But as we show in the next section the conjecture is trivially false in that case. It is not too difficult to find in ZFC examples (Example 2.1) of AEC that have no model above $\aleph_1$ but that are $\aleph_1$-categorical [She09a, Bal09]. And in $L_{\omega_1,\omega}(Q)$, it is consistent (via Martin's axiom) that there are $\aleph_1$-categorical sentences with no model of cardinality greater than $2^{\aleph_0}$. But those sentences have many models in $2^{\aleph_0}$. In this note we describe how for $L_{\omega_1,\omega}$, there are major advances on the target problem. They use extensions of ZFC but rather mild ones; the initials below refer to the 'Weak Continuum Hypothesis' and the 'Very Weak Continuum Hypothesis':

WGCH: Cardinal exponentiation is an increasing function.

VWGCH: Cardinal exponentiation is an increasing function below $\aleph_\omega$.

This leaves us with two more precise questions.

1. Does the proof of the conjecture for $L_{\omega_1,\omega}$ (see Section 4) really need VWGCH?
2. Is the conjecture 'eventually true' for AEC's[3]?

Much of core mathematics studies either properties of particular structures of size at most the continuum or makes assertions that are totally cardinal independent. E.g., if every element of a group has order two then the group is abelian. Model theory and even more clearly infinitary model theory allows the investigation of 'structural properties' that are cardinal dependent such as: existence of models, spectra of stability, and number of models and existence of decompositions. Often these properties can be tied to global conditions such as the existence of a 'good' notion of dependence.

2. Abstract Elementary Classes

We begin by discussing the notion of an abstract elementary class. The examples show that this is too broad a class to be 'reasonable' for our target problem. But

[2]The main gap theorem, every first order theory either eventually has the maximal number of models or the number of models is bounded by a small function, has the same flavor. And in fact the argument for this result arose after Shelah's consideration of the infinitary problems.

[3]For much positive work in this direction see [She09a].

some positive results can be proved in this general setting; this generality exposes more clearly what is needed for the argument by avoiding dependence on accidental syntactic features.

An abstract elementary class $(\boldsymbol{K}, \prec_{\boldsymbol{K}})$[4] is a collection of structures for a fixed vocabulary τ that satisfy, where $A \prec_{\boldsymbol{K}} B$ means in particular A is a substructure of B,

1. If $A, B, C \in \boldsymbol{K}$, $A \prec_{\boldsymbol{K}} C$, $B \prec_{\boldsymbol{K}} C$ and $A \subseteq B$ then $A \prec_{\boldsymbol{K}} B$;
2. Closure under direct limits of $\prec_{\boldsymbol{K}}$-embeddings;
3. Downward Löwenheim-Skolem. If $A \subset B$ and $B \in \boldsymbol{K}$ there is an A' with $A \subseteq A' \prec_{\boldsymbol{K}} B$ and $|A'| \leq |A| + \mathrm{LS}(\boldsymbol{K})$.

The invariant $\mathrm{LS}(\boldsymbol{K})$, is a crucial property of the class. The class of well-orderings satisfies the other axioms (under end extension) but is not an AEC.

Two easy examples are: First order and $L_{\omega_1,\omega}$-classes; $L(Q)$ classes have Löwenheim-Skolem number $\aleph_1$. For the second case one has to be careful about the definition of $\prec_{\boldsymbol{K}}$ – being an $L(Q)$-elementary submodel does not work (a union of a chain can make $(Qx)\phi(x)$ become true even if it is false along the chain).

The notion of AEC has been reinterpreted in terms of category theory by Kirby: "Abstract Elementary Categories" [Kir08] and by Lieberman: "AECs as accessible categories" [Lie].

It is easy to see that just AEC is too weak a condition for the general conjecture.

Example 2.1 The class of well-orderings with order-type at most ω_1 with $\prec_{\boldsymbol{K}}$ as initial segment is an AEC with $\aleph_1$ countable models. It is $\aleph_1$-categorical and satisfies both amalgamation and joint embedding but is not ω-Galois stable [Kue08]. And in fact there is no model of cardinality $\aleph_2$. So this universe is neither wide nor deep.

Let's clarify the specific meaning of the amalgamation property in this context. The arrows here denote morphisms in the abstract elementary class; various strengthening requiring certain maps to be inclusions are well-known.

Definition 2.2 *The class* $\boldsymbol{K}$ *satisfies the* amalgamation property *if for any situation with* $A, M, N \in \boldsymbol{K}$*:*

[4]Naturally we require that both $\boldsymbol{K}$ and $\prec_{\boldsymbol{K}}$ are closed under isomorphism.

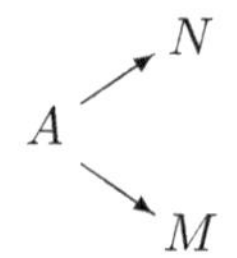

there exists an $N_1 \in \boldsymbol{K}$ such that

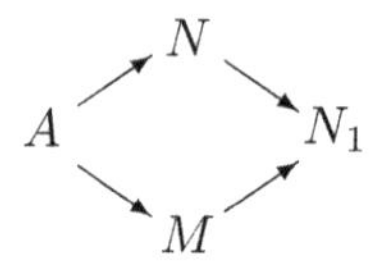

Note that we have required the base structure A to be in $\boldsymbol{K}$; this is sometimes referred to as 'model amalgamation'. Requiring amalgamation over arbitrary substructures A is a much stronger condition, which fails for important natural examples such as Zilber's pseudo-exponential field [Zil04]. There is much work in homogenous model theory where the stronger homogeneity condition is assumed.

The existence of amalgamations is an absolutely fundamental problem for AEC and for any study of infinitary logic. In first order logic it is easy to show that for complete theories amalgamation always holds over models with $\prec$ as elementary extension. And it holds over arbitrary subsets of models if T admits elimination of quantifiers. Here is a basic example of failure for a complete sentence of $L_{\omega_1,\omega}$.

Example 2.3 Let T be the first order theory in a language with binary relation symbols $\langle E_i : i < \omega \rangle$ that asserts the E_i are infinitely many refining equivalence relations with binary splitting.

Using $L_{\omega_1,\omega}$ the equivalence relation E_∞, the intersection of the given equivalence relations, is definable. Add two unary predicates (blue and red) and the infinitary axioms

1. Each E_∞-class contains infinitely many elements.
2. Every element of an E_∞-class is red or every element is blue.
3. Blue and red divide the E_∞-classes into dense and codense subsets of the natural linear order of the paths.

Now it is easy to check that these axioms are $\aleph_0$-categorical but fail amalgamation (since a new path may be either red or blue).

We introduced the notion of abstract elementary class in this paper in order to state One Completely General Result which can be found in I.3.8 of [She09a] or [She83b, Bal09].

Theorem 2.4 (WGCH) *Suppose $\lambda \geq \mathrm{LS}(\boldsymbol{K})$ and $\boldsymbol{K}$ is λ-categorical. If amalgamation fails in λ there are 2^{λ^+} models in $\boldsymbol{K}$ of cardinality λ^+.*

As opposed to many other results in the study of abstract elementary classes which rely on an additional collection of model theoretic hypotheses, this result is about *all* AEC's. Moreover, variants of the proposition recur repeatedly in the proof of the main result being expounded. The argument uses weak diamond and is primarily combinatorial; it proceeds directly from the definition of an AEC. The result fails under $MA + \neg CH$. An example is presented in [She87, She09a] and a simpler one in [Bal09]. It is an AEC (even given by a theory in $L(Q)$) which fails amalgamation in $\aleph_0$, but becomes $\aleph_1$-categorical in a forcing extension. But it remains open whether there are such examples in $L_{\omega_1,\omega}$. Easy examples ([BKS09]) show the categoricity is a necessary condition for Theorem 2.4. This has a fundamental impact on the structure of the main proof. Because of this we must pass to complete sentences and gain categoricity in $\aleph_0$. One strategy in Shelah's approach through frames in [She09a] evades the categoricity difficulty by restricting to subclasses of the AEC, e.g. the λ-saturated models.

Amalgamation plays a fundamental role in the study of AECs. One line of research pioneered by Shelah [She99] and highly developed by Grossberg, VanDieren, and Lessmann in a series of papers (e.g. [GV06]) assumes arbitrarily large models, joint embedding, and amalgamation; under strong model theoretic assumptions the results are proved in ZFC. An account of this work with full references to the published papers appears in Part II of [Bal09]. In this paper we focus on earlier work on $L_{\omega_1,\omega}$, which is a little more concrete as the logic is fixed. But it is more general in another way. Rather than assuming amalgamation, failure of amalgamation is shown to create width. Both amalgamation and the existence of large models are proved for narrow classes; this brings the set theoretic difficulties into view. The work of Hyttinen and Kesala on finitary AEC (e.g. [HK07]) continues the program of assuming arbitrarily large models and amalgamation. But, even stronger model theoretic assumptions lead to the development of a geometric stability theory. Several further directions of study in AEC are explored in [She09a]. The introduction to that book surveys the field and explains Shelah's viewpoint. The method of frames, expounded in [She09a], provides an approach to the problem of building larger models from categoricity in one or several successive uncountable cardinals; he attempts to avoid the traces of compactness that simplify the work starting at $\aleph_0$ and $\aleph_1$ in $L_{\omega_1,\omega}$. In other papers Shelah (e.g. [She01]) considers the general problem of eventual categoricity assuming large cardinal axioms.

3. From $L_{\omega_1,\omega}$ to First Order

We begin by translating the problem from infinitary logic into the study of specific subclasses of models of first order theories. This removes the distraction of developing new notions of each syntactic idea (e.g. type) for each fragment of $L_{\omega_1,\omega}$. More subtly, for technical reasons we need to restrict to complete sentences in $L_{\omega_1,\omega}$. (This restriction to complete sentences is automatic in the first order case but its legitimacy is only proved in certain cases for infinitary logic).

Definition 3.1 *For Δ a fragment of $L_{\omega_1,\omega}$, a Δ-theory T is Δ-complete if for every Δ-sentence ϕ, $T \models \phi$ or $T \models \neg\phi$. We may write* complete *when $\Delta = L_{\omega_1,\omega}$.*

Definition 3.2

1. *A model M of a first order theory is called* atomic *if each finite sequence from M realizes a principal type over the empty set – one generated by a single formula.*
2. *An* atomic class *is an aec, consisting of the atomic models of a complete first order theory T with elementary submodel as the notion of strong submodel. $\mathbb{M}$ is a large saturated model of T; it is usually not atomic. A set $A \subset \mathbb{M}$ is an* atomic set *if each finite sequence from A realizes a principal type over the empty set -generated by a single formula.*

The study of categoricity (at least from $\aleph_1$ upwards), in $L_{\omega_1,\omega}$ can be translated to the study of atomic models of a first order theory. This is non-trivial. The argument begins with a fundamental result from the early 60's.

Theorem 3.3 (Chang/Lopez-Escobar) *Let ϕ be a sentence in $L_{\omega_1,\omega}$ in a countable vocabulary τ. Then there is a countable vocabulary τ' extending τ, a first order τ'-theory T, and a countable collection of τ'-types Γ such that reduct is a* 1-1 *map from the models of T which omit Γ onto the models of ϕ.*

The proof is straightforward. E.g., for any formula ψ of the form $\bigwedge_{i<\omega} \phi_i$, add to the language a new predicate symbol $R_\psi(\mathbf{x})$. The theory T will contain the sentences for each subformula ψ of ϕ:

$$(\forall \mathbf{x})[R_\psi(\mathbf{x}) \rightarrow \phi_i(\mathbf{x})]$$

for $i < \omega$ and omit the type $p = \{\neg R_\psi(\mathbf{x})\} \cup \{\phi_i : i < \omega\}$. There are similar requirements for other steps in the inductive definition of θ.

Thus we have restricted to the models of a theory that omit a family Γ of types, but that may realize some non-principal types. Shelah observed that if T had only countably many types then a similar expansion of the vocabulary gives a T' such that the required interpretation is obtained by omitting *all* non-principal types.

That is, the object of study is the atomic models of T'. This further reduction is technically important. In particular it implies ω-categoricity.

But why can we assume that the T associated with θ has only countably many types over the empty set? We need a few definitions to give an explanation.

Definition 3.4 *Fix a sentence $\phi \in L_{\omega_1,\omega}$ and let Δ be a countable fragment of $L_{\omega_1,\omega}$ containing ϕ.*

1. *A τ-structure M is Δ-small if M realizes only countably many Δ-types (over the empty set).*
2. *An $L_{\omega_1,\omega}$-sentence ϕ is Δ-small if there is a countable set X of complete Δ-types over the empty set and each model realizes some subset of X.*

'small' means $\Delta = L_{\omega_1,\omega}$

It is easy to see that if M is small then M satisfies a complete sentence. If ϕ is small then Scott's argument for countable models generalizes and there is a complete sentence ψ_ϕ such that: $\phi \wedge \psi_\phi$ has a countable model. So ψ_ϕ implies ϕ. But ψ_ϕ is not in general unique. For example ϕ might be just the axioms for algebraically closed fields. Two choices for ψ_ϕ are the Scott sentence of the prime field and the Scott sentence for the model of transcendence degree $\aleph_0$. Only the second has an uncountable model.

We can make an appropriate choice of ψ_ϕ if ϕ is $\aleph_1$-categorical. There are two ingredients in the choice.

Theorem 3.5 (Shelah) *If ϕ has an uncountable model M that is Δ-small for every* countable *Δ and ϕ is $\aleph_1$-categorical then ϕ is implied by a complete sentence ψ with a model of cardinality $\aleph_1$.*

This result appears first in [She83a]. It is retold in [Bal09]; in [Bal07], we adapt the argument to give a model theoretic proof of a result of Makkai (obtained by admissible set theory) that a counterexample to Vaught's conjecture is not $\aleph_1$-categorical. The crux of Shelah's argument is an appeal to the non-definability of well-order in $L_{\omega_1,\omega}$.

The second step is to require that for each countable fragment Δ there are only countably many Δ-types over the empty set. If ϕ has arbitrarily large models this is easy by using Ehrenfeucht-Mostowski models. But if not, the only known argument is from few models in $\aleph_1$ and depends on a subtle argument of Keisler [Kei71] (See also Appendix C of [Bal09].)

Theorem 3.6 (Keisler) *If ϕ has $< 2^{\aleph_1}$ models of cardinality $\aleph_1$, then each model of ϕ is Δ-small for every* countable *Δ.*

Now Theorems 3.5 and 3.6 immediately yield.

Theorem 3.7 *[Shelah] If ϕ has $< 2^{\aleph_1}$ models of cardinality $\aleph_1$, then there is a complete sentence ψ such that ψ implies ϕ and ψ has an uncountable model. In particular, if ϕ is $\aleph_1$-categorical there is a Scott sentence for the model in $\aleph_1$, i.e. the model in $\aleph_1$ is small. So an atomic class $\boldsymbol{K}$ is associated with ϕ .*

It is easy to construct a sentence ϕ such that no completion has an uncountable model, i.e. no uncountable model is small. Let τ contain binary relations E_n for $n < \omega$. Let ϕ assert that the E_n are refining equivalence relations with binary splitting. And that there do not exist two distinct points that are E_n equivalent for all n. And add a countable set A of constants that realize a dense set of paths. Now every uncountable model realizes uncountably many distinct types over A.

We have the following question, which is open if $\kappa > \aleph_1$.

Question 3.8 *If ϕ is κ-categorical must the model in κ be small?*

Thus for technical work we will consider the class of atomic models of first order theories. Our notion of type will be the usual first order one - but we must define a restricted Stone space.

Definition 3.9 *Let A be an atomic set; $S_{\mathrm{at}}(A)$ is the collection of $p \in S(A)$ such that if $\boldsymbol{a} \in \mathbb{M}$ realizes p, $A\boldsymbol{a}$ is atomic.*

Here $\mathbb{M}$ is the monster model for the ambient theory T; in interesting cases it is not atomic. And the existence[5] of a monster model for the atomic class associated with a sentence categorical in some set of cardinals is a major project. (It follows from excellence. After Theorem 4.3, we see under VWGCH categoricity up to $\aleph_\omega$ is sufficient).

Definition 3.10 $\boldsymbol{K}$ *is λ-stable if for every* model M *in* $\boldsymbol{K}$ *(thus necessarily atomic) with cardinality λ, $|S_{at}(M)| = \lambda$.*

The insistence that M be a model is essential. The interesting examples of pseudo-exponential fields, covers of Abelian varieties and the basic examples of Marcus and Julia Knight all are ω-stable but have countable sets A with $|S_{\mathrm{at}}(A)| > \aleph_0$.

With somewhat more difficulty than the first order case, one obtains:

[5] The difficulties we discuss here concern obtaining amalgamation. For simplicity, think only of gaining a monster model in λ with $\lambda^{<\lambda} = \lambda$. Weakening that hypothesis is a different project (See [Bal09, Hod93]) or any first order stability book for comments on the cardinality question.)

Theorem 3.11 *For a class $\boldsymbol{K}$ of atomic models, ω-stable implies stable in κ for all κ.*

A fundamental result in model theory is Morley's proof that an $\aleph_1$-categorical first order theory is ω-stable. This argument depends on the compactness theorem in a number of ways. The key idea is to construct an Ehrenfeucht-Mostowski model over a well-order of cardinality $\aleph_1$. Such a model realizes only countably many types over any countable submodel. But the existence of the model depends on a compactness argument in the proof of the Ehrenfeucht-Mostowski theorem. Further, this only contradicts ω-stability because amalgamation allows the construction from a model M_0 in $\aleph_0$ that has uncountably many types over it an elementary extension M_1 of M_0 with power $\aleph_1$ that realizes all of them. And again amalgamation in the first order case is a consequence of compactness. In $L_{\omega_1,\omega}$, the work of Keisler and Shelah evades the use of compactness – but at the cost of set theoretic hypotheses.

Theorem 3.12 (Keisler-Shelah) *Let $\boldsymbol{K}$ be the atomic models of a countable first order theory. If $\boldsymbol{K}$ is $\aleph_1$-categorical and $2^{\aleph_0} < 2^{\aleph_1}$ then $\boldsymbol{K}$ is ω-stable.*

This proof uses WCH directly and weak diamond via 'The Only Completely General Result'. That is, from amalgamation failure of ω-stability yields a model of cardinality $\aleph_1$ that realizes uncountably many types from $S_{\mathrm{at}}(M)$ for a countable model M. Naming the elements of M yields a theory which has uncountably many types over the empty set. Thus by Theorem 3.6 the new theory has $2^{\aleph_1}$ models in $\aleph_1$ and (since $2^{\aleph_0} < 2^{\aleph_1}$) so does the original theory. Is CH is necessary?

Example 3.13 *There are examples [She, Bal09], of an AEC $\boldsymbol{K}$ and even one given by a sentence of $L_{\omega_1,\omega}(Q)$ such that MA + $\neg$ CH imply $\boldsymbol{K}$ is $\aleph_1$-categorical but $\boldsymbol{K}$*

a) is not ω-stable

b) does not satisfy amalgamation even for countable models.

Laskowski (unpublished) showed the example proposed for $L_{\omega_1,\omega}$ by Shelah [She87, She09a] fails. The question of whether such an $L_{\omega_1,\omega}$-example exists is a specific strategy for answering the next question.

Question 3.14 *Is categoricity in $\aleph_1$ of a sentence of $L_{\omega_1,\omega}$ absolute (with respect suitable forcings)?*

By suitable, I mean that, e.g., it is natural to demand cardinal preserving. This result has resisted a number of attempts although as we lay out in Section 5, many other fundamental notions of the model theory of $L_{\omega_1,\omega}$ are absolute.

4. The Conjecture for $L_{\omega_1,\omega}$

Using the notion of splitting, a nice theory of independence (Definition 5.6) can be defined for ω-stable atomic classes [She83a, She83b, Bal09]. This allows the formulation of the crucial notion of excellence and the proof of a version of Morley's theorem. We won't discuss the details but outline some aspects of the argument. These results are non-trivial but the exposition of the entire situation in [Bal09] occupies less than 100 pages.

The concept of an independent system of models is hard to grasp although it is playing an increasing role in many areas of model theory. Rather than repeating the notation heavy definition (see [She83b, Bal09, Les05] or various first order stability texts.), I give a simple example. Let X be a set of n algebraically independent elements in an algebraically closed field. For each $Y \subsetneq X$, let M_Y be the algebraic closure of Y. The M_Y form a independent system of $2^n - 1$-models. This is exactly the concept needed in Zilber's theory of quasiminimal excellence. For Shelah's more general approach the notion is axiomatized using the independence notion from the previous paragraph. In the example, there is clearly a prime model over the union of the independent system. In various more complicated algebraic examples (e.g. [BZ00]) the existence of such a prime model is non-trivial. Here we discuss how to find one from model theoretic hypotheses.

Definition 4.1 *Let* $\boldsymbol{K}$ *be an atomic class.* $\boldsymbol{K}$ *is* excellent *if* $\boldsymbol{K}$ *is* ω*-stable and any of the following equivalent conditions hold.*

For any finite independent system of countable models with union C*:*

1. $S_{at}(C)$ *is countable.*
2. *There is a unique primary model over* C*.*
3. *The isolated types are dense in* $S_{at}(C)$*.*

The key point is that this is a condition of 'n-dimensional amalgamation'. A primary model is a particulary strong way of choosing a prime model over C. Thus, condition ii) specifies the existence of a strong kind of amalgamation of n independent models. This definition emphasizes the contrast of the current situation with first order logic; condition 1) does *not* follow from ω-stability. See [Bal09] for details of the notation.

Note that excellence is a condition on countable models. It has the following consequence for models in *all* cardinalities. The key to this extension is the proof that n-dimensional amalgamation in $\aleph_n$ implies $n-1$-dimensional amalgamation in $\aleph_{n+1}$. Thus amalgamation for all n in $\aleph_0$ implies amalgamation for all n below $\aleph_\omega$ and then for all cardinals by a short argument.

Theorem 4.2 (Shelah (ZFC)) *If an atomic class $\boldsymbol{K}$ is excellent and has an uncountable model then*

1. *$\boldsymbol{K}$ has models of arbitrarily large cardinality;*
2. *Categoricity in one uncountable power implies categoricity[6] in all uncountable powers.*

This result is in ZFC but extensions of set theory are used to obtain excellence. Recall that by VWGCH we mean the assertion: $2^{\aleph_n} < 2^{\aleph_{n+1}}$ for $n < \omega$. The following is an immediate corollary of Theorem 4.6.

Theorem 4.3 (Shelah (VWGCH)) *An atomic class $\boldsymbol{K}$ that is categorical in $\aleph_n$ for each $n < \omega$ is excellent.*

We remarked after Definition 3.9 on the difficulty of constructing a monster model for an atomic class associated with a sentence categorical in some power. Of course such a monster model in appropriate cardinalities is immediate from the amalgamation property. But, even assuming categoricity up to $\aleph_\omega$, we need to use the VWGCH to get excellence, then derive amalgamation and finally a monster model.

The requirement of categoricity below $\aleph_\omega$ in Theorem 4.3 is essential. Baldwin-Kolesnikov [BK09] (refining [HS90]) show:

Theorem 4.4 *For each $2 \leq k < \omega$ there is an $L_{\omega_1,\omega}$-sentence ϕ_k such that:*

1. *ϕ_k has an atomic model in every cardinal.*
2. *ϕ_k is categorical in μ if $\mu \leq \aleph_{k-2}$;*
3. *ϕ_k is not categorical in any μ with $\mu > \aleph_{k-2}$;*
4. *ϕ_k has the (disjoint) amalgamation property;*

Note that of course the ϕ_k are not excellent. There is one further refinement on the 'wide' vrs 'deep' metaphor. How wide?

Definition 4.5 *We say*

1. *$\boldsymbol{K}$ has* few *models in power λ if $I(\boldsymbol{K}, \lambda) < 2^\lambda$.*

[6] In contrast to some authors, we say $\boldsymbol{K}$ is categorical in κ if there is *exactly* one model in cardinality κ.

2. $\boldsymbol{K}$ *has* very few *models in power* $\aleph_n$ *if* $I(\boldsymbol{K}, \aleph_n) \leq 2^{\aleph_{n-1}}$.

These are equivalent under GCH. And Shelah argues on the last couple of pages of [She83b] (see also [She0x]) that they are equivalent under $\neg 0^+$. But in general we have a theorem and a conjecture [She83a, She83b], which differ only in the word 'very'.

Theorem 4.6 (Shelah) *(For $n < \omega$, $2^{\aleph_n} < 2^{\aleph_{n+1}}$.) An atomic class $\boldsymbol{K}$ that has at least one uncountable model and that has* very *few models in $\aleph_n$ for each $n < \omega$ is excellent.*

Conjecture 4.7 (Shelah) *(For $n < \omega$, $2^{\aleph_n} < 2^{\aleph_{n+1}}$.) An atomic class $\boldsymbol{K}$ that has at least one uncountable model and that has* few *models in $\aleph_n$ for each $n < \omega$ is excellent.*

The proof of Theorem 4.6 uses the technology of atomic classes very heavily. But the calculation of the categoricity spectrum in Theorem 4.2.2 can be lifted to arbitrary sentences of $L_{\omega_1,\omega}$ by a calculation [She83a, She83b], reported as Theorem 25.19 of [Bal09].

5. Absoluteness of Properties of Atomic Classes

As remarked in the introduction, one of the significant attributes of first order stability theory is that the basic notions: stable, ω-stable, superstable, $\aleph_1$-categoricity can be seen absolute in very strong ways. We sketch proofs of similar results, except the open $\aleph_1$-categoricity, for $L_{\omega_1,\omega}$. This section and the appendix tie together some results which are folklore with the use of well-known methods which are systematically applied to discuss the case of $L_{\omega_1,\omega}$. We are indebted for discussions with Alf Dolich, Paul Larson, Chris Laskowski, and Dave Marker for clarifying the issues. Among the few places model theoretic absoluteness issues have recently been addressed in print is [She09b]. Earlier accounts include [Sac72, Bar75].

For example a first order theory T is unstable just if there is a formula $\phi(\mathbf{x}, \mathbf{y})$ such for every n

$$T \models (\exists \mathbf{x}_1, \ldots \mathbf{x}_n \exists \mathbf{y}_1, \ldots \mathbf{y}_n) \bigwedge_{i<j} \phi(\mathbf{x}_i, \mathbf{y}_j) \wedge \bigwedge_{i \geq j} \neg\phi(\mathbf{x}_i, \mathbf{y}_j)$$

This is an arithmetic statement and so is absolute by basic properties of absoluteness [Kun80, Jec87]. In first order logic, ω-stability is Π^1_1; there is no consistent tree[7] $\{\phi_i^{\sigma(i)}(x_\sigma, \boldsymbol{a}_\sigma \restriction n) : \sigma \in 2^\omega, i < \omega\}$. With a heavier use of effective

[7] We use the convention that $\phi^{\sigma(i)}\phi(x)$ denotes $\phi(x)$ or $\neg\phi(x)$ depending on whether $\sigma(i)$ is 0 or 1.

descriptive set theory, suggested by Dave Marker, the same applies for the atomic class case.

To demonstrate absoluteness of various concepts of infinitary logic we need the full strength of the Shoenfield absoluteness lemma. In this section, we work with *atomic classes*, Definition 3.2. We noted Shelah's observation Theorem 3.7 that each $\aleph_1$-categorical sentence of $L_{\omega_1,\omega}$ determines such a class. In this section we first show absoluteness for various properties of atomic classes. In the last theorem, we show that the properties for sentences of $L_{\omega_1,\omega}$ remain absolute although in some cases they are more complex. The Appendix (written by David Marker) makes a precise definition of a formula in $L_{\omega_1,\omega}$ as a subset of $\omega^{<\omega}$ so that we can apply descriptive set theoretic techniques. It gives an effective analysis of the transformation in Theorem 3.3. The appendix fixes some notation for the rest of the paper and clarifies the complexity of a number of basic notions; e.g., that the collection of complete sentences in $L_{\omega_1,\omega}$ is complete Π^1_1.

Theorem 5.1 (Shoenfield absoluteness Lemma) *If*

1. *$V \subset V'$ are models of ZF with the same ordinals and*
2. *ϕ is a lightface Π^1_2 predicate of a set of natural numbers*

then for any $A \subset N$, $V \models \phi(A)$ iff $V' \models \phi(A)$.

Note that this trivially gives the same absoluteness results for Σ^1_2-predicates.

Lemma 5.2 (Atomic models)

1. *'T has an atomic model' is an arithmetic property of T.*
2. *'M is an atomic model of T' is an arithmetic property of M and T.*
3. *For any vocabulary τ, the class of countable atomic τ-structures, M, is Borel.*

Proof. The first condition is given by: for every formula $\phi(\mathbf{x})$ there is a $\psi(\mathbf{x})$, consistent with T, such that $\psi(\mathbf{x}) \rightarrow \phi(\mathbf{x})$ and for every $\chi(\mathbf{x})$, $\psi(\mathbf{x}) \rightarrow \chi(\mathbf{x})$ or $\psi(\mathbf{x}) \rightarrow \neg\chi(\mathbf{x})$. Let $\theta(M,T)$ be the arithmetic predicate of the reals M,T asserting that T is the theory of M. Now the third condition is a Δ^1_1-predicate of M given by: there exists (for all) T such that $\theta(M,T)$ and for every $\boldsymbol{a} \in M$, there exists a T-atom $\psi(\mathbf{x})$ such that $M \models \psi(\boldsymbol{a})$. $\square_{5.2}$

Earlier versions of this paper had weaker characterizations; e.g, a Σ^1_2 characterization of ω-stability and Π^1_2 characterization of excellence. Marker pointed out the application of Harrison's theorem, Fact 5.4.ii, to improve the result to Π^1_1.

Definition 5.3 $x \in \omega^\omega$ *is* hyperarithmetic *if* $x \in \Delta^1_1$. x *is* hyperarithmetic *in* y, *written* $x \leq_{\rm hyp} y$, *if* $x \in \Delta^1_1(y)$.

Fact 5.4 1. *The predicate* $\{(x, y) : x \leq_{hyp} y\}$ *is* Π^1_1.

2. *If* $K \subset \omega^\omega$ *is* Σ^1_1, *then for any* y, K *contains an element which is not hyperarithmetic in* y *if and only if* K *contains a perfect set.*

The unrelativized version of statement 1) is II.1.4.ii of [Sac90]; the relativized version is 7.15 of [Mar]. Again, the unrelativized version of statement 2) is III.6.2 of [Sac90]; in this case the relativization is routine. $\square_{5.4}$

In the next theorem, the atomic set A must be regarded as an element of ω^ω. There are at least two ways to think of this: 1) a pair (M, A) where is M is a countable atomic model of T and A is a subset (automatically atomic) of M or 2) as a pair (A, Φ) where Φ is the diagram of A as an atomic subset of the monster model $\mathbb{M}$.

Lemma 5.5 (Marker) *Let* $\boldsymbol{K}$ *be an atomic class (Definition 3.2) with a countable complete first order theory* T.

1. *Let* A *be a countable atomic set. The predicate of* p *and* A, *'*p *is in* $S_{\rm at}(A)$*', is arithmetic.*

2. *'*$S_{\rm at}(A)$ *is countable' is a* Π^1_1*-predicate of* A.

Proof. i) Note first that '$q(\mathbf{x})$ is a principal type over $\emptyset$ in T' is an arithmetic property. Now p is in $S_{\rm at}(A)$ if and only if for all $\boldsymbol{a} \in A$, $p \restriction \boldsymbol{a}$ is a principal type. So this is also arithmetic.

ii) By i), the set of p such that 'p is in $S_{\rm at}(A)$' is arithmetic (*a fortiori* Σ^1_1) in A, so by Lemma 5.4.ii, each such p is hyperarithmetic in A. Since the continuum hypothesis holds for Σ^1_1-sets, '$S_{\rm at}(A)$ is countable' is formalized by:

$$(\forall p)[p \in S_{\rm at}(A) \rightarrow (p \leq_{\rm hyp} A)],$$

which is Π^1_1. $\square_{5.5}$

In order to show the absoluteness of excellence we need some more detail on the notion of independence. We will use item i) of Definition 4.1. The independent families of models [She83b, Bal09] in that definition are indexed by subsets of n with strictly less than n elements; we denote this partial order by $\mathcal{P}^-(n)$. We will show that independence of models is an arithmetic property.

Definition 5.6 1. *A complete type* p *over* A splits *over* $B \subset A$ *if there are* $\mathbf{b}, \mathbf{c} \in A$ *which realize the same type over* B *and a formula* $\phi(\mathbf{x}, \mathbf{y})$ *with* $\phi(\mathbf{x}, \mathbf{b}) \in p$ *and* $\neg\phi(\mathbf{x}, \mathbf{c}) \in p$.

2. *Let ABC be atomic. We write $A \underset{C}{\perp} B$ and say A is free or independent from B over C if for any finite sequence $\mathbf{a}$ from A, $\mathrm{tp}(\mathbf{a}/B)$ does not split over some finite subset of C.*

Lemma 5.7 *Let T be a complete countable first order theory. The properties that the class of atomic models of T is*

1. *ω-stable*
2. *excellent*

are each given by a Π^1_1-formula of set theory and so are absolute.

Proof. 1) The class of atomic models of T is ω-stable if and only if for every atomic model M, '$S_{\mathrm{at}}(M)$ is countable'. This property is Π^1_1 by Lemma 5.5.

2) The class of atomic models of T is excellent if and only if for any finite set of countable atomic models $\{A_s : s \in \mathcal{P}^-(n)\}$ that form an independent system, with $A = \bigcup\{A_s : s \in \mathcal{P}^-(n)\}$, $S_{\mathrm{at}}(A)$ is countable. Here we have universal quantifiers over finite sequences of models (using a pairing function, this is quantifying over a single real). The stipulation that the diagram is independent requires repeated use of the statement $A \underset{C}{\perp} B$, where A, B, C are finite unions of the models in the independent system. This requires quantification over finite sequences from the A_s; thus, it is arithmetic. The assertion '$S_{\mathrm{at}}(A)$ is countable' is again π^1_1 by Lemma 5.5 and we finish. $\square_{5.7}$

Lemma 5.8 *The property that an atomic class $\boldsymbol{K}$ has arbitrarily large models is absolute. In fact it is Σ^1_1.*

Proof. Let $\boldsymbol{K}$ be the class of atomic models of a first order theory T in a vocabulary τ. $\boldsymbol{K}$ has arbitrarily large models if and only there are $\hat{T}$, $\hat{\tau}$, M and C such that $\hat{T}$ is a Skolemization of T in a vocabulary $\hat{\tau}$ and M is a countable model of $\hat{T}$ such that $M \restriction \tau$ is atomic and M contains an infinite set C of $\hat{\tau}$-indiscernibles. This formula is Σ^1_1. $\square_{8.8}$

Finally, following Lessmann [Les05, Bal09], we prove that the absolute 'Baldwin-Lachlan'-characterization of first order $\aleph_1$-categoricity has a natural translation to the $L_{\omega_1,\omega}$ situation; the resulting property of atomic classes is absolute and in ZFC it implies $\aleph_1$-categoricity. But we do not see how to derive it from $\aleph_1$-categoricity without using the Continuum hypothesis. We need some definitions. To be a bit more specific we speak of Vaughtian triples instead of Vaughtian pairs.

Definition 5.9 *The formula $\phi(x, \mathbf{c})$ with $\mathbf{c} \in M \in \boldsymbol{K}$, is* big *if for any $M' \supseteq A$ with $M' \in \boldsymbol{K}$ there exists an N' with $M' \prec_{\boldsymbol{K}} N'$ and with a realization of $\phi(x, \mathbf{c})$ in $N' - M'$.*

This definition has no requirements on the cardinality of M, M', N' so it is saying that $\phi(\mathbf{x}, \mathbf{c})$ has as many solutions as the size of the largest models in $\boldsymbol{K}$. This condition is equivalent to one on countable models. A translation of Lemma 25.2 of [Bal09] gives:

Lemma 5.10 *Let $A \subseteq M$ and $\phi(x, \mathbf{c})$ be over A. The following are equivalent.*

1. *There is an N with $M \prec N$ and $c \in N - M$ satisfying $\phi(x, \mathbf{c})$;*
2. *$\phi(x, \mathbf{c})$ is big.*

The significance of this remark is that it makes '$\phi(x, \mathbf{c})$ is big' a Σ^1_1 predicate.

Definition 5.11
1. *A triple (M, N, ϕ) where $M \prec N \in \boldsymbol{K}$ with $M \neq N$, ϕ defined over M, ϕ big, and $\phi(M) = \phi(N)$ is called a* Vaughtian triple.
2. *We say $\boldsymbol{K}$ admits (κ, λ), witnessed by ϕ, if there is a model $N \in \boldsymbol{K}$ with $|N| = \kappa$ and $|\phi(N)| = \lambda$ and ϕ is big.*

Now we have the partial characterization.

Lemma 5.12 *Let $\boldsymbol{K}$ be a class of atomic models. If $\boldsymbol{K}$ is ω-stable and has no Vaughtian triples then $\boldsymbol{K}$ is $\aleph_1$-categorical. The hypothesis of this statement is Π^1_1.*

Proof. The sufficiciency of the condition is found by tracing results in [Bal09]: ω-stability gives the existence of a quasiminimal formula ϕ. Note from the proof of Theorem 24.1 in [Bal09] that ω-stability is sufficient to show that there are prime models over independent subsets of cardinality $\aleph_1$. (The point of excellence is that higher dimensional amalgamation is needed to extend this result to larger sets.) So if $|M| = \aleph_1$, there is an $N \prec_{\boldsymbol{K}} M$ which is prime over a basis for $\phi(M)$. As noted in Chapter 2 of [Bal09], this determines N up to isomorphism (again without use of excellence because we are in $\aleph_1$). So we are done unless $N \precneqq M$. But then Löwenheim-Skolem gives us a countable Vaughtian triple, contrary to hypothesis. $\square_{5.12}$

Since the second condition below is true if $2^{\aleph_0} < 2^{\aleph_1}$ and we have shown the conclusion of that condition is absolute, we have:

Corollary 5.13 *$\aleph_1$-categoricity is absolute for atomic classes if and only if in ZFC $\aleph_1$-categoricity implies countable amalgamation and ω-stabity.*

Consequence 5.14 *Let $\boldsymbol{K}$ be a class of atomic models. $\aleph_1$-categoricity of $\boldsymbol{K}$ is absolute between models of set theory that satisfy either of the following conditions.*

1. *$\boldsymbol{K}$ has arbitrarily large members and $\boldsymbol{K}$ has amalgamation in $\aleph_0$, or*
2. $2^{\aleph_0} < 2^{\aleph_1}$.

Proof. Each hypothesis implies the characterization in Lemma 5.12. $\square_{5.15}$

Note, the hypothesis of condition 1) is absolute. It seems unlikely that $\aleph_1$-categoricity implies the existence of arbitrarily large models in $\boldsymbol{K}$; but no counterexample has yet been constructed. The use of the continuum hypothesis is central to current proofs that $\aleph_1$-categoricity implies amalgamation and ω-stability. For general AEC, Example 3.13 shows ZFC does not imply the assertion A): $\aleph_1$-categoricity implies amalgamation in $\aleph_0$ and ω-stability. But [FK0x] have shown (employing standard forcings) that for each AEC $\boldsymbol{K}$ that fails amalgamation in $\aleph_0$, there is a model of set theory such that in that model $2^{\aleph_0} = 2^{\aleph_1}$, $\boldsymbol{K}$ continues to fail amalgamation in $\aleph_0$, and $\boldsymbol{K}$ has $2^{\aleph_1}$ models in $\aleph_1$. So assertion A) does not imply CH.

Consequence 5.15 *Let $\boldsymbol{K}$ be a class of atomic models. Categoricity in all cardinals is absolute between models of set theory that satisfy the VWGCH.*

Proof. Under VWGCH, categoricity in all powers is equivalent to the Π^1_1-condition: excellence with no two cardinal models. $\square_{5.15}$

Theorem 5.16 *Each of the properties that a complete sentence of $L_{\omega_1,\omega}$ is ω-stable, excellent, or has no two-cardinal models is Σ^1_2.*

Proof. Let $Q(T)$ denote any of the conditions above as a property of the first order theory T in a vocabulary τ^*. Now write the following properties of the complete sentence ϕ in vocabulary τ.

1. ϕ is a complete sentence.
2. There exists a $\tau^* \supseteq \tau$ and τ^* theory T satisfying the following.
   (a) T is a complete theory that has an atomic model.
   (b) The reduct to τ of any atomic model of T satisfies ϕ.

(c) There is a model M of ϕ and there exists an expansion of M to an atomic model of T.

(d) $Q(T)$.

Proof. We know condition 1) is Π^1_1. Condition 2) is an existential function quantifier followed by conditions which are at worst Π^1_1. $\square_{5.16}$

So, as far as we know the conditions on sentences of $L_{\omega_1,\omega}$ are more complicated than those for atomic classes and the application of Harrison's lemma[8] was needed to obtain absoluteness of these conditions for sentences of $L_{\omega_1,\omega}$.

6. Complexity

We prove the following claim. This result was developed in conversation with Martin Koerwien and Sy Friedman at the CRM Barcelona and benefitted from further discussion with Dave Marker.

Claim 6.1 *The class of countable models whose automorphism groups admit a complete left invariant metric is* Π^1_1 *but not* Σ^1_1.

Our proof is by propositional logic from known results of Gao [Gao96] and Deissler [Dei77].

Definition 6.2 *A countable model is* minimal *(equivalently* non-extendible*) if it has no proper* $L_{\omega_1,\omega}$*-elementary submodel.*

We showed in Lemma 5.2 that the class of atomic structures is Borel. The following claim is an easy back and forth.

Claim 6.3 *If* M *is atomic,* τ*-elementary submodel is the same as* $L_{\omega_1,\omega}(\tau)$*-elementary submodel.*

Claim 6.3 shows an atomic model is minimal iff it is minimal in first order logic. Note that the class of first order minimal models is obviously Π^1_1. Now if the class of minimal models were Borel, it would follow that the class of minimal atomic (equal first order minimal prime) models is also Borel. But Corollary 2.6 of Deissler [Dei77] asserts for first order theories:

Lemma 6.4 (Deissler) *There is a countable relational vocabulary* τ *such that the class of minimal prime models for* τ *is not* Σ^1_1.

---

[8]Grossberg has pointed out that by suitably modifying the rank for ω-stable atomic classes the result could be given a direct model theoretic proof. This is slightly tricky because this rank will only be defined on some atomic sets.

Gao [Gao96] characterized non-extendible models in terms of metrics on their automorphism group.

Lemma 6.5 (Gao) *The following are equivalent:*

1. *Aut(M) admits a compatible left-invariant complete metric.*
2. *There is no $L_{\omega_1,\omega}$-elementary embedding from M into itself which is not onto.*

So we can transfer to the characterization of automorphism groups and prove Claim 6.1.

Gao pointed out to me that Malicki [Mal10] recently proved a related result: the class of Polish groups with a complete left invariant metric) is $\mathbf{\Pi}_1^1$ but not $\mathbf{\Sigma}_1^1$. We now analyze the connection between the two results and show the properties studied are Borel equivalent. This observation was made jointly with Christian Rosendal.

Recall that S_∞ is is a Borel subspace of N^N . $\mathbb{S}\mathbb{G}(S_\infty)$ denotes the collection of closed subgroups of S_∞. It is contained in $\mathbb{F}$, the hyperspace of closed subsets of S_∞. $\mathbb{F}$ is a standard Borel space with the Effros-Borel structure generated by

$$\{F \in \mathbb{F} : F \cap U \neq \emptyset\}$$

for some open $U \subset S_\infty$. Proposition 1 of [Mal10] implies that with this topology $\mathbb{S}\mathbb{G}(S_\infty)$ is a standard Borel space.

Claim 6.6 *The map A taking M to $AutM$ mapping the standard Borel space of countable atomic models models into $\mathbb{S}\mathbb{G}(S_\infty)$ is Borel.*

Proof. We have to show that for any basic open set $X \in \mathbb{S}\mathbb{G}(S_\infty)$, $A^{-1}(X)$ is a Borel subset of $\mathcal{A}$. That is, for fixed open U, if X is the set of F with $F \cap U \neq \emptyset$, the inverse image of X is Borel in the space of atomic models.

Say U is all permutations mapping $\boldsymbol{a}$ to b where $\boldsymbol{a}, \mathrm{b} \in N^n$ Now there is a $g \in \mathrm{aut(M)}$ mapping $\boldsymbol{a}$ to b if and only if $\boldsymbol{a}$ and b realize the same type in M if and only if they satisfy the same formulas over the empty set, which is a Borel condition. We have shown.

Corollary 6.7 *The class of Polish groups with a complete left invariant metric is Π_1^1 but not Σ_1^1.*

Conversely, we want to reduce the CLI groups to the class of minimal atomic models. The reduction is a map B from a group G acting on N to a structure M on N with $\mathrm{aut(M)} = \mathrm{G}$. This is easily done by mapping G to a structure with universe N which has a predicate for each orbit of G on N^n.

Deissler also uses a vocabulary with infinitely many n-ary predicates for each n so the vocabulary is in fact the same for both directions of reduction.

7. Conclusion

The spectrum problem for first order theories motivated many technical developments that eventually had significant algebraic consequences. A similar possibility for application of infinitary logic to algebraic problems is suggested by Zilber's program [Zil06, Zil04]. But the basic development is far more difficult and less advanced. The notion of excellence provides one useful context. And others are being developed under the guise of abstract elementary classes and metric abstract elementary classes. But while first order stability theory is developed in ZFC, the current development of the model theory of $L_{\omega_1,\omega}$ uses a (rather weak) extension of set theory: the VWGCH. This raises both model theoretic and set theoretic questions. The proof of the 'one completely general result', Theorem 2.4, is a fundamentally combinatorial argument using no sophisticated model theoretic lemmas. The current proof uses $2^\lambda < 2^{\lambda^+}$. Can this hypothesis be removed?

Like first order logic such fundamental definitions of $L_{\omega_1,\omega}$ as satisfaction, ω-stablity, and excellence are absolute. And in fact the complexity of their description can often be computed. But while $\aleph_1$-categoricity is seen (by a model theoretic argument) to be absolute in the first order case, this issue remains open for $L_{\omega_1,\omega}$.

We have also investigated the complexity of various properties of $L_{\omega_1,\omega}$-sentences and associated atomic classes. It is shown in Lemma 8.7 that the graph of the translation from a sentence to a finite diagram (T, Γ) is arithmetic. In Theorem 5.16, we avoided a precise calculation of the translation from a complete sentence to the atomic models of a first order theory. The tools of the appendix should allow a careful computation of this complexity. Note that while, for example, we showed ω-stability was Π^1_1 as a property of an atomic class, we only showed it to be Σ^1_2 as a property of the $L_{\omega_1,\omega}$-sentence.

8. Appendix: Basic definability notions for $L_{\omega_1,\omega}$ by David Marker

Fix a vocabulary τ and let $\mathbb{X}_\tau$ be the Polish space of countable τ-structures with universe ω. Our first goal is to describe the collection of codes for $L_{\omega_1,\omega}(\tau)$-formulas. This is analogous to the construction of Borel codes in descriptive set theory.

Definition 8.1 *1.* A labeled tree *is a non-empty tree $T \subseteq \omega^{<\omega}$ with functions l and v with domain T such that for any $\sigma \in T$ one of the following holds:*

• *σ is a terminal node of T then $l(\sigma) = \psi$ where ψ is an atomic τ-formula and $v(\sigma)$ is the set of free variables in ψ;*
• *$l(\sigma) = \neg$, $\sigma\hat{}0$ is the unique successor of σ in T and $v(\sigma) = v(\sigma\hat{}0)$;*
• *$l(\sigma) = \exists v_i$, $\sigma\hat{}0$ is the unique successor of σ in T and $v(\sigma) = v(\sigma\hat{}0) \setminus \{i\}$;*
• *$l(\sigma) = \bigwedge$ and $v(\sigma) = \bigcup_{\sigma\hat{}i \in T} v(\sigma\hat{}i)$ is finite.*

2. *A* formula *ϕ is a well founded labeled tree (T, l, v). A* sentence *is a formula where $v(\emptyset) = \emptyset$.*

Proposition 8.2 *The set of labeled trees is arithmetic. The set of formulas is complete-Π^1_1, as is the set of sentences.*

Now it is easy to see:

Proposition 8.3 *There is $R(x, y) \in \Pi^1_1$ and $S(x, y) \in \Sigma^1_1$ such that if ϕ is a sentence and $M \in \mathbb{X}_\tau$, then*

$$M \models \phi \Leftrightarrow R(M, \phi) \Leftrightarrow S(M, \phi).$$

In particular, $\{(M, \phi) : \phi$ is a sentence and $M \models \phi\}$ is Π^1_1, but for any fixed ϕ, $\mathrm{Mod}(\phi) = \{M \in \mathbb{X}_\tau : M \models \phi\}$ is Borel, indeed $\Delta^1_1(\phi)$.

Proof. We define a predicate 'f is a *truth definition* for the labeled tree (T, l, v) in M' as follows.

• The domain of f is the set of pairs (σ, μ) where $\sigma \in T$ and $\mu : v(\sigma) \to M$ is an assignment of the free variables at node σ and $f(\sigma, \mu) \in \{0, 1\}$.

• If $l(\sigma) = \psi$ an atomic formula, then $f(\sigma, \mu) = 1$ if and only if ψ is true in M when we use μ to assign the free variables.

• If $l(\sigma) = \neg$, then $f(\sigma, \mu) = 1$ if and only if $f(\sigma\hat{}0, \mu) = 0$.

• If $l(\sigma) = \exists v_i$ there are two cases. If $v_i \in v(\sigma\hat{}0)$, then $f(\sigma, \mu) = 1$ if and only if there is $a \in M$ such that $f(\sigma\hat{}0, \mu^*) = 1$, where $\mu^* \supset \mu$ is the assignment where $\mu^*(v_i) = a$. Otherwise, $f(\sigma, \mu) = f(\sigma\hat{}0, \mu)$.

• If $l(\sigma) = \bigwedge$, then $f(\sigma, \mu) = 1$ if and only if $f(\sigma\hat{}i, \mu|v)(\sigma\hat{}i) = 1$ for all i such that $\sigma\hat{}i \in T$.

This predicate is arithmetic. If ϕ is a sentence, there is a unique truth definition f for ϕ in M. Let

$R(x, y) \Leftrightarrow x \in \mathbb{X}_\tau$ and y is a labeled tree and $f(\emptyset, \emptyset) = 1$ for all truth definitions f for y in x

and

$S(x, y) \Leftrightarrow y$ is a labeled tree and there is a truth definition f for y in x such that $f(\emptyset, \emptyset) = 1$. $\square_{8.3}$

Notation 8.4 *We write that a property of a set of reals is $\Pi^1_1 \wedge \Sigma^1_1$ if it is defined by the conjunction of a Π^1_1 and a Σ^1_1 formula.*

Proposition 8.5 *$\{\phi : \phi$ is a satisfiable sentence$\}$ is $\Pi^1_1 \wedge \Sigma^1_1$, but neither $\mathbf{\Pi}^1_1$ nor $\mathbf{\Sigma}^1_1$.*

Proof. 'ϕ is a sentence' is Π^1_1; 'there is a model for ϕ' is equivalent to $\exists x\, S(x, \phi)$ which is Σ^1_1.

The set of satisfiable sentences is not $\mathbf{\Sigma}^1_1$ since otherwise the set of underlying trees would be a $\mathbf{\Sigma}^1_1$-set of trees and there would be a countable bound (e.g. Theorem 3.12 of [MW85]), on their heights.

We show that the set of satisfiable sentences is not $\mathbf{\Pi}^1_1$ by constructing a reduction of non-well ordered linear orders to satisfiable sentences.

Let $\tau = \{U, V, <, s, f, 0, c_n : n \in \omega\}$

For each linear order $\prec$ of ω we write down an $L_{\omega_1,\omega}$ sentence $\phi_\prec$ asserting:

- the universe is the disjoint union of U and V;
- $U = \{c_0, c_1, \ldots\}$ all of which are distinct;
- $<$ is a linear order of U;
- $c_n < c_m$, if $n \prec m$;
- s is a successor function on V and $V = \{0, s(0), s(s(0)), \ldots\}$;
- $f : V \to U$ and $f(s(n)) < f(n)$ for all n.

It $\prec$ is not a well order, and $n_0 \succ n_1 \succ \ldots$ is an infinite descending chain, then by defining $f(n) = c_n$ we get a model of $\phi_\prec$. On the other hand if $\prec$ is a well order we can find no model of $\phi_\prec$.

Thus $\prec \mapsto \phi_\prec$ is a reduction of non-well-ordered linear orders to $\{\phi : \phi$ is satisfiable$\}$ which is impossible if satisfiability is $\mathbf{\Pi}^1_1$. $\square_{8.5}$

We now effectivize Chang's observation (Lemma 3.3) that for each sentence ϕ in $L_{\omega_1,\omega}$ we can find a first order theory T^* in a vocabulary τ^* and a countable set Γ of partial τ^*-types such that the models of ϕ are exactly the τ-reducts of models of T^* that omit all the types in Γ.

Definition 8.6 *A* Chang-assignment *to a labeled tree (T, l, v) is a pair of functions S, γ with domain T such that $S(\sigma)$ is a set of sentences in the vocabulary $\tau_\sigma = \tau \cup \{R_\tau : \tau \supseteq \sigma\}$, where τ and σ are in T and R_τ is a relation symbol in $|v(\tau)|$-variables and $\gamma(\sigma)$ is a function with domain ω such that each $\gamma(\sigma)(n)$ is a partial τ_σ type.[9] We also require:*

- *if $l(\sigma) = \psi$ is atomic, $S(\sigma) = \{\forall \overline{v}(\sigma)(R_\sigma(\overline{v}) \leftrightarrow \psi\}$, and each $\gamma(\sigma)(i) = \{v_1 \neq v_1\}$;*

[9] We allow relation symbols in 0 variables, but these could easily be eliminated.

• *if* $l(\sigma) = \neg$, *then* $S(\sigma) = S(\sigma\hat{}0) \cup \{\forall\overline{v}(\sigma)R_\sigma \leftrightarrow NegR_{\sigma\hat{}0}\}$ *and* $\gamma(\sigma) = \gamma(\sigma\hat{}0)$.

• *if* $l(\sigma) = \exists v_i$, *then* $S(\sigma) = S(\sigma\hat{}0) \cup \{\forall\overline{v}(\sigma)R_\sigma \leftrightarrow \exists v_i R_{\sigma\hat{}0}\}$ *and* $\gamma(\sigma) = \gamma(\sigma\hat{}0)$.

• *if* $l(\sigma) = \bigwedge$; *then* $S(\sigma) = \bigcup_{\sigma\hat{}i \in T} S(\sigma\hat{}i) \cup \{\forall\overline{v}(\sigma)(R_\sigma \to R_{\sigma\hat{}i}) : \sigma\hat{}i \in T\}$. *Fix* $\mu : \omega \times \omega \to \omega$ *be a pairing function. Let*

$$\gamma(\sigma)(0) = \{R_\sigma, \neg R_{\sigma\hat{}i} : \sigma\hat{}i \in T\}$$

and

$$\gamma(\sigma)(\mu(i,n)+1) = \begin{cases} \gamma(\sigma\hat{}i)(n) & \text{if } \sigma\hat{}i \in T \\ \{v_1 \neq v_1\} & \text{otherwise.} \end{cases}$$

In other words $\gamma(\sigma)$ *lists all the types listed by the successors of* σ *and the additional type* $\{R_\sigma, \neg R_{\sigma\hat{}i} : \sigma\hat{}i \in T\}$.

It is now easy to see:

Lemma 8.7 *The predicate "*(S,γ) *is a Chang-assignment for the labeled tree* (T,l,v)*" is arithmetic. If* ϕ *is a sentence then there is a unique Chang-assignment for* ϕ.

To simplify notation we will call (T,Γ) the Chang-assignment, where T is the theory $S(\emptyset)$ and Γ is the set of types $\gamma(\emptyset)(0), \gamma(\emptyset)(1), \ldots$.

The following remark is implicit in [GS86].

Lemma 8.8 *The property that a sentence* ϕ *of* $L_{\omega_1,\omega}$ *has arbitrarily large models is absolute. In fact it is* $\Pi^1_1 \wedge \Sigma^1_1$, *but neither* $\mathbf{\Pi}^1_1$ *nor* $\mathbf{\Sigma}^1_1$.

Proof. A τ-sentence ϕ has arbitrarily large models if and only if there is a Chang-assignment (T,Γ), $\tau^* \supseteq \tau$ and $T^* \supseteq T$ a Skolemized τ^*-theory such that there is a model of T^* omitting all types in Γ and containing an infinite set of τ^*-indiscernibles. This condition is Σ^1_1 once we restrict to the Π^1_1-set of sentences.

For any sentence ϕ let ϕ^* be the sentence which asserts we have two sorts, the first of which is a model of ϕ and the second is an infinite set with no structure. Then ϕ is satisfiable if and only if ϕ^* has arbitrarily large models. Thus $\phi \mapsto \phi^*$ is a reduction of satisfiable sentences to sentences with arbitrarily large models. By Proposition 8.5, the set of sentences with arbitrarily large models is neither $\mathbf{\Sigma}^1_1$ nor $\mathbf{\Pi}^1_1$. $\square_{8.8}$

Recall that an $L_{\omega_1,\omega}$-sentence is *complete* if and only if it is satisfiable and any two countable models are isomorphic. This is easily seen to be Π^1_2. Drawing on some results of Nadel, we show that in fact:

Theorem 8.9 $\{\phi\colon\phi$ *is a complete sentence*$\}$ *is complete-*Π^1_1.

The argument requires some preparation. We begin by recalling the usual Karp-Scott back-and-forth analysis.

Definition 8.10 *If M and N are τ-structures, we inductively define $\sim_\alpha$, by:*

$(M, \boldsymbol{a}) \sim_0 (N, \mathbf{b})$ if $M \models \phi(\boldsymbol{a})$ if and only if $N \models \phi(\mathbf{b})$ for all atomic τ-formulas ϕ.

For all ordinals α, $(M, \boldsymbol{a}) \sim_{\alpha+1} (N, \mathbf{b})$ if for all $c \in M$ there is $d \in N$ such that $(M, \boldsymbol{a}, c) \sim_\alpha (N, \mathbf{b}, d)$ and for all $d \in N$ there is $c \in M$ such that $(M, \boldsymbol{a}, c) \sim_\alpha (N, \mathbf{b}, d)$

For all limit ordinals β, $(M, \boldsymbol{a}) \sim_\beta (N, \mathbf{b})$ if and only if $(M, \boldsymbol{a}) \sim_\alpha (N, \mathbf{b})$ for all $\alpha < \beta$.

A classical fact is that $(M, \boldsymbol{a}) \sim_\alpha (N, \mathbf{b})$ if and only if $M \models \phi(\boldsymbol{a}) \Leftrightarrow N \models \phi(\mathbf{b})$ for all formulas ϕ of quantifier rank at most α.

We say that ϕ has *Scott rank* α if α is the least ordinal such that if $M, N \models \phi$ and $(M, \boldsymbol{a}) \sim_\alpha (N, \mathbf{b})$ then $(M, \boldsymbol{a}) \sim_\beta (N, \mathbf{b})$ for all ordinals β.

We need to analyze the complexity of $\sim_\alpha$.

Definition 8.11 *Let* WO* *(the class of pseudo-well-orders) be the set of all linear orders R with domain ω such that:*

i) 0 is the R-least element;

ii) if n is not R-maximal, then there is y such that xRy and there is no z such that xRz and zRx, we say y is the R-successor of x and write $y = s_R(x)$. If $n \neq 0$ is not an R-successor we say it is an R-limit.

Note that WO*, $s_R(n) = m$ and 'n is an R-limit' are arithmetic.

Definition 8.12 *We say that z is an R*-analysis of *M* and *N if*

i) $z \subseteq \omega \times \bigcup_{n\in\omega}(\omega^n \times \omega^n)$;

ii) $(0, \boldsymbol{a}, \mathbf{b}) \in z$ if and only if $M \models \phi(\boldsymbol{a}) \leftrightarrow N \models \phi(\mathbf{b})$ for all quantifier free ϕ;

iii) if $(n, \boldsymbol{a}, \mathbf{b})$ and mRn, then $(m, \boldsymbol{a}, \mathbf{b})$;

iv) $(s_R(n), \boldsymbol{a}, \mathbf{b}) \in z$ if and only if for all $c \in \omega$ there is $d \in \omega$ such that $(n, \boldsymbol{a}\hat{\ }c, \mathbf{b}\hat{\ }d) \in z$ and for all $d \in \omega$ there is $c \in \omega$ such that $(n, \boldsymbol{a}\hat{\ }c, \mathbf{b}\hat{\ }d) \in z$;

v) if n is an R-limit, then $(n, \boldsymbol{a}, \mathbf{b}) \in z$ if and only if $(m, \boldsymbol{a}, \mathbf{b}) \in z$ for all mRn.

Note:

- $\{(z, R, M, N)$: 'z is an R-analysis'$\}$ is arithmetic.

• Suppose R is a well-order of order type α. Let $\beta(n) < \alpha$ be the order type of $\{m : mRn\}$. If z is an R-analysis of M, N, then

$$(n, \boldsymbol{a}, \mathbf{b}) \in z \text{ if and only if } (M, \boldsymbol{a}) \sim_{\beta(n)} (N, \mathbf{b}).$$

In particular, there is a unique R-analysis of M, N.

We need two results of Mark Nadel.

Theorem 8.13 (Nadel) *a) If ϕ is complete, then there is $M \models \phi$ with $M \leq_{\rm hyp} \phi$.*

b) If ϕ is complete then the Scott rank of ϕ is at most $\mathrm{qr}(\phi) + \omega$ *where* $\mathrm{qr}(\phi)$ *is the quantifier rank of ϕ.*

a) is [Nad74b] Theorem 2, while b) is [Nad74a] Theorem 5.1. For completeness we sketch the proofs.

a) Add new constants $c_1, c_2, \ldots$ to τ. Let F be a countable fragment such that $\phi \in F$, we can choose F arithmetic in ϕ. Let $S = \{s : s$ a finite set of F-sentences using only finitely many c_i such that $\phi \models \exists \overline{v} \bigwedge_{\psi \in s} \psi(\overline{v})\}$. S is a consistency property. Since ϕ is complete,

$$\phi \models \theta \Leftrightarrow \forall M\,(M \models \phi \rightarrow M \models \theta) \Leftrightarrow \exists M\,(\mathcal{M} \models \phi \wedge M \models \theta).$$

It follows that S is $\Delta^1_1(\phi, F)$ and hence $S \leq_{\rm hyp} \phi$. Using the consistency property S one can easily construct $M \models \phi$ with $M \leq_{\rm hyp} \phi$.

b) Let F be as above. Since ϕ is complete, there are only countably many F-types. By the Omitting Types Theorem for $L_{\omega_1,\omega}$, there is a model of ϕ where every element satisfies an F-complete formula. Since ϕ is complete, this is true in the unique countable model M.

The usual arguments show that we can do a back and forth in M with F-types. Thus if $\boldsymbol{a}, \mathbf{b}b$ in $\mathcal{M}$ and $(M, \boldsymbol{a}) \equiv_F (M, \mathbf{b})$ then there is an automorphism of M mapping $\boldsymbol{a}$ to $\mathbf{b}$. If we pick α such that every ψ is F has quantifier rank below α and $(M, \boldsymbol{a}) \sim_\alpha (M, \mathbf{b})$, then $(M, \boldsymbol{a}) \sim_\beta (M, \mathbf{b})$ for all β. Thus the Scott rank of ϕ is at most α.

If F is the smallest fragment containing ϕ, every formula in F has Scott rank below $\mathrm{qr}(\phi) + \omega$, so this is an upper bound on the Scott rank. $\square_{8.13}$

Proof of Theorem 8.9. First note that if α is a bound on the Scott rank of models of ϕ, then any two countable models M and N of ϕ are isomorphic if and only if we can do a back-and forth construction using $\sim_\alpha$. Thus by Nadel's Theorems, a sentence ϕ is complete if and only if

i) $(\exists M) M \leq_{\rm hyp} \phi \wedge M \models \phi$ and

ii) $\exists \alpha$ recursive in ϕ such that for all $M, N \models \phi$ if $\boldsymbol{a} \in M, \mathbf{b} \in N$ and $(M, \boldsymbol{a}) \sim_\alpha (N, \mathbf{b})$, then for all $c \in M$ there is $d \in \overline{N}$ such that $(M, \boldsymbol{a}, c) \sim_\alpha (N, \mathbf{b}, d)$.

i) is easily seen to be Π^1_1, using Fact 5.4.

ii) is equivalent to $\forall M, N \models \phi\ (\exists R, \exists z) z \leq_{\text{hyp}} \phi\ , R \in WO^*$ and z is an R-analysis of M and N and there is an n such that if $\boldsymbol{a}, c \in M, \mathbf{b} \in N$ with $(n, \boldsymbol{a}, \mathbf{b}) \in z$, then there is $d \in N$ such that $(n, \boldsymbol{a}, c, \mathbf{b}, d) \in z$. This is also Π^1_1, again using Fact 5.4.

Finally, to each linear order $\prec$ of ω we will assign an $L_{\omega_1,\omega}$ sentence $\phi_\prec$ such that $\prec$ is a well order if and only if $\phi_\prec$ is complete. This will show that $\{\phi : \phi$ is complete$\}$ is $\boldsymbol{\Pi}^1_1$-complete.

The vocabulary τ is $\{P_n : n \in \omega\}$ where P_n is a unary predicate.

• We say that every element is in some P_n.

• We say that each P_n is infinite and that if $n \prec m$, then $P_n \subset P_m$ and $P_m \setminus P_n$ is infinite.

• Moreover if $\forall m \prec n \exists k\ m \prec k \prec n$, then we also say that $P_n \setminus \bigcup_{m \prec n} P_m$ is infinite.

If $\prec$ is a well ordering, then $\phi_\prec$ is $\aleph_0$-categorical as for each n we just put $\aleph_0$ elements in each $P_n \setminus \bigcup_{m \prec n} P_m$.

On the other hand if $n_0 \succ n_1 \succ \ldots$ is an infinite descending chain let $X = \{m : m \prec n_i$ for all $i\}$. We can put any number of elements in

$$\bigcap_{i=0}^{\infty} P_{n_i} \setminus \bigcup_{m \in X} P_m,$$

so $\phi_\prec$ is not complete. $\square_{8.9}$

References

[Bal07] J.T. Baldwin. Vaught's conjecture, do uncountable models count? *Notre Dame Journal of Formal Logic*, pages 1–14, 2007.

[Bal09] John T. Baldwin. *Categoricity*. Number 51 in University Lecture Notes. American Mathematical Society, 2009. www.math.uic.edu/~ jbaldwin.

[Bar75] J. Barwise, editor. *Admissible sets and structures*. Perspectives in Mathematical Logic. Springer-Verlag, 1975.

[BK09] John T. Baldwin and Alexei Kolesnikov. Categoricity, amalgamation, and tameness. *Israel Journal of Mathematics*, 170, 2009. also at `www.math.uic.edu/\~\jbaldwin`.

[BKS09] J.T. Baldwin, A. Kolesnikov, and S. Shelah. The amalgamation spectrum. *Journal of Symbolic Logic*, 74:914–928, 2009.

[BZ00] M. Bays and B.I. Zilber. Covers of multiplicative groups of an algebraically closed field of arbitrary characteristic. preprint: arXive math.AC/0401301, 200?

[CK73] C.C. Chang and H.J Keisler. *Model theory*. North-Holland, 1973. 3rd edition 1990.

[Dei77] Rainer Deissler. Minimal models. *The Journal of Symbolic Logic*, 42:254–260, 1977.

[FK0x] Sy-David Friedman and Martin Koerwien. On absoluteness of categoricity in aecs. preprint, 200x.

[Gao96] Su Gao. On automorphism groups of countable structures. *The Journal of Symbolic Logic*, 63:891–896, 1996.

[GS86] R. Grossberg and Saharon Shelah. On the number of non isomorphic models of an infinitary theory which has the order property part A. *Journal of Symbolic Logic*, 51:302–322, 1986.

[GV06] R. Grossberg and M. VanDieren. Categoricity from one successor cardinal in tame abstract elementary classes. *The Journal of Mathematical Logic*, 6:181–201, 2006.

[HK07] T. Hyttinen and M. Kesälä. Superstability in simple finitary AECs. *Fund. Math.*, 195(3):221–268, 2007.

[Hod93] W. Hodges. *Model Theory*. Cambridge University Press, 1993.

[HS90] Bradd Hart and Saharon Shelah. Categoricity over P for first order T or categoricity for $\phi \in l_{\omega_1\omega}$ can stop at $\aleph_k$ while holding for $\aleph_0, \cdots, \aleph_{k-1}$. *Israel Journal of Mathematics*, 70:219–235, 1990.

[Jec87] T. Jech. *Multiple Forcing*, volume 88 of *Cambridge Topics in Mathematics*. Cambridge University Press, 1987.

[Kei71] H.J Keisler. *Model theory for Infinitary Logic*. North-Holland, 1971.

[Kir08] Jonathan Kirby. Abstract elementary categories. `http://arxiv.org/abs/1006.0894v1`, 2008.

[Kue08] D. W. Kueker. Abstract elementary classes and infinitary logics. *Annals of Pure and Applied Logic*, pages 274–286, 2008.

[Kun80] K. Kunen. *Set Theory, An Introduction to Independence Proofs*. North Holland, 1980.

[Les05] Olivier Lessmann. An introduction to excellent classes. In Yi Zhang, editor, *Logic and its Applications*, Contemporary Mathematics, pages 231–261. American Mathematical Society, 2005.

[Lie] M. Lieberman. Accessible categories vrs aecs. preprint:`www.math.lsa.umich.edu/~liebermm/vita.html`.

[Mal10] M. Malicki. On Polish groups admitting a complete left-invariant metric. *Journal of Symbolic Logic*, 2010. to appear: "Journal of Symbolic Logic".

[Mar] D. Marker. Descriptive set theory. Notes from 2002; `http://www.math.uic.edu/~marker/math512/dst.pdf`.

[MW85] R. Mansfield and G. Weitkamp. *Recursive Aspects of Descriptive Set Theory*. Oxford University Press, 1985.

[Nad74a] Mark Nadel. More Lowenheim-Skolem results for admissible sets. *Israel J. Math.*, 18:53–64, 1974.

[Nad74b] Mark Nadel. Scott sentences and admissible sets. *Ann. Math. Logic*, 7:267–294, 1974.

[Sac72] G. Sacks. *Saturated Model Theory*. Benjamin, Reading, Mass., 1972.

[Sac90] G. Sacks. *Higher Recursion Theory*. Springer-Verlag, Berlin Heidelberg, 1990.

[She] Saharon Shelah. Abstract elementary classes near $\aleph_1$ sh88r. revision of Classification of nonelementary classes II, Abstract elementary classes; on the Shelah archive.

[She75] S. Shelah. Categoricity in $\aleph_1$ of sentences in $L_{\omega_1,\omega}(Q)$. *Israel Journal of Mathematics*, 20:127–148, 1975. paper 48.

[She83a] S. Shelah. Classification theory for nonelementary classes. I. the number of uncountable models of $\psi \in L_{\omega_1\omega}$ part A. *Israel Journal of Mathematics*, 46:3:212–240, 1983. paper 87a.

[She83b] S. Shelah. Classification theory for nonelementary classes. I. the number of uncountable models of $\psi \in L_{\omega_1\omega}$ part B. *Israel Journal of Mathematics*, 46;3:241–271, 1983. paper 87b.

[She87] Saharon Shelah. Classification of nonelementary classes II, abstract elementary classes. In J.T. Baldwin, editor, *Classification theory (Chicago, IL, 1985)*, pages 419–497. Springer, Berlin, 1987. paper 88: Proceedings of the USA–Israel Conference on Classification Theory, Chicago, December 1985; volume 1292 of *Lecture Notes in Mathematics*.

[She99] S. Shelah. Categoricity for abstract classes with amalgamation. *Annals of Pure and Applied Logic*, 98:261–294, 1999. paper 394. Consult Shelah for post-publication revisions.

[She01] Saharon Shelah. Categoricity of theories in $L_{\kappa\omega}$, when κ is a measureable cardinal, part II. *Fundamenta Mathematica*, 170:165–196, 2001.

[She09a] S. Shelah. *Classification Theory for Abstract Elementary Classes*. Studies in Logic. College Publications `www.collegepublications.co.uk`, 2009. Binds together papers 88r, 600, 705, 734 with introduction E53.

[She09b] S. Shelah. Model theory without choice? categoricity. *Journal of Symbolic Logic*, 74:361–401, 2009.

[She0x] S. Shelah. Non-structure in λ^{++} using instances of WGCH. paper 838, 200x.

[Zil04] B.I. Zilber. Pseudo-exponentiation on algebraically closed fields of characteristic 0. *Annals of Pure and Applied Logic*, 132:67–95, 2004.

[Zil06] B.I. Zilber. Covers of the multiplicative group of an algebraically closed field of characteristic 0. *J. London Math. Soc.*, pages 41–58, 2006.

K-TRIVIALS ARE NEVER CONTINUOUSLY RANDOM

George Barmpalias
Institute for Logic, Language and Computation, Universiteit van Amsterdam,
P.O. Box 94242, 1090 GE Amsterdam, The Netherlands
barmpalias@gmail.com
http://www.barmpalias.net

Noam Greenberg
School of Mathematics, Statistics and Operations Research
Victoria University
Wellington, New Zealand
greenberg@msor.vuw.ac.nz

Antonio Montalbán
Department of Mathematics
University of Chicago
5734 S. University ave.
Chicago, IL 60637, USA
antonio@math.uchicago.edu

Theodore A. Slaman
Department of Mathematics
University of California, Berkeley
Berkeley, CA 94720-3840 USA
slaman@math.berkeley.edu

1. Introduction

In [RS07, RS08], Reimann and Slaman raise the question "For which infinite binary sequences X do there exist continuous probability measures μ such that X is effectively random relative to μ?". They defined the collection NCR_1 of binary sequences for which such measures do not exist (we give formal definitions below), and showed, for example, that NCR_1 is countable, indeed that every sequence in NCR_1 is hyperarithmetic. In this paper we contribute toward the understanding of NCR_1 by showing that it contains all sets which are Turing reducible to an incomplete, recursively enumerable set. In particular, NCR_1 contains all K-trivial sets.

1.1. *Randomness relative to continuous measures*

We begin by reviewing the basic definitions needed to precisely formulate this question.

Notation 1.1.

- For $\sigma \in 2^{<\omega}$, $[\sigma]$ is the basic open subset of 2^ω consisting of those X's which extend σ. Similarly, for W a subset of $2^{<\omega}$, let $[W]$ be the open set given by the union of the basic open sets $[\sigma]$ such that $\sigma \in W$.
- For $U \subseteq 2^\omega$, $\lambda(U)$ denotes the measure of U under the uniform distribution. Thus, $\lambda([\sigma])$ is $1/2^\ell$, where ℓ is the length of σ.

Definition 1.2. A *representation* m of a probability measure μ on 2^ω provides, for each $\sigma \in 2^{<\omega}$, a sequence of intervals with rational endpoints, each interval containing $\mu([\sigma])$, and with lengths converging monotonically to 0.

Definition 1.3. Suppose that $Z \in 2^\omega$. A *test relative to* Z, or Z*-test*, is a set $W \subseteq \omega \times 2^{<\omega}$ which is recursively enumerable in Z. For $X \in 2^\omega$, X *passes* a test W if and only if there is an n such that $X \notin [W_n]$.

Definition 1.4. Suppose that m represents the measure μ on 2^ω and that W is an m-test.

- W is *correct for* μ if and only if for all n, $\mu([W_n]) \leq 2^{-n}$.
- W is *Solovay-correct for* μ if and only if $\sum_{n\in\omega} \mu([W_n]) < \infty$.

Definition 1.5. $X \in 2^\omega$ is 1*-random relative to a representation* m *of* μ if and only if X passes every m-test which is correct for μ. When m is understood, we say that X is 1-random relative to μ.

By an argument of Solovay, see [Nie09], X is 1-random relative to a representation m of μ if an only if for every m-test which is Solovay-correct for μ, there are infinitely many n such that $X \notin [W_n]$.

Definition 1.6. $X \in \mathrm{NCR}_1$ if and only if there is no representation m of a continuous measure μ such that X is 1-random relative to the representation m of μ.

In [RS08], Reimann and Slaman show that if X is not hyperarithmetic, then there is a continuous measure μ such that X is 1-random relative to μ. Conversely, Kjøs-Hanssen and Montalbán, see [Mon05], have shown that if X is an element of a countable Π^0_1-class, then there is no continuous measure for which X is 1-random. As the Turing degrees of the elements of countable Π^0_1-classes are cofinal in the Turing degrees of the hyperarithmetic sets, the smallest ideal in the Turing degrees that contains the degrees represented in NCR_1 is exactly the Turing degrees of the hyperarithmetic sets.

In [RSte], Reimann and Slaman pose the problem to find a natural Π^1_1-norm for NCR_1 and to understand its connection with the natural norm mapping a hyperarithmetic set X to the ordinal at which X is first constructed. As of the writing of this paper, this problem is open in general, but completed in [RSte] for $X \in \Delta^0_2$.

Suppose that $X \in \Delta^0_2$ and that for all n, $X(n) = \lim_{t\to\infty} X_t(n)$, where $X_t(n)$ is a computable function of n and t. Let g_X be the convergence function for this approximation, that is for all n, $g_X(n)$ is the least s such that for all $t \geq s$ and all $m \leq n$, $X_t(m) = X(m)$. Let f_X be function obtained by iterated application of g_X: $f_X(0) = g_X(0)$ and $f_X(n+1) = g_X(f_X(n))$.

For a representation m of a continuous measure μ, the granularity function s_m maps $n \in \omega$ to the least ℓ found in the representation of μ by m such that for all σ of length ℓ, $\mu([\sigma]) < 1/2^n$. Note that, s_m is well-defined by the compactness of 2^ω.

Theorem 1.7 (Reimann and Slaman [RSte]). *Let X be a Δ^0_2 set and let f_X be the function defined as above. If X is 1-random relative the representation m of μ, then the granularity function s_m for μ is eventually bounded by f_X.*

In the proof of Theorem 1.7, the possibility that s_m eventually bounds f_X is ruled out since it would imply that X is recursive in m, contradicting X's being 1-random. The possibility that neither function bound the other is ruled out by the direct construction of a Martin-Löf test for μ, defined using g and the recursive approximation to X, which X would fail, again contradicting X's being 1-random.

It follows that, for Δ^0_2 sets X, there is a continuous measure relative to which X is 1-random if and only if there is a continuous measure whose granularity is eventually bounded by f_X. The latter condition is arithmetic, again by a compactness argument.

1.2. *K-triviality*

K-triviality is a property of sequences which characterizes another aspect of their being far from random. We briefly review this notion and the results surrounding it. A full treatment is given in Nies [Nie09].

For $\sigma \in 2^{<\omega}$, let $K(\sigma)$ denote the prefix-free Kolmogorov complexity of σ. Intuitively, given a universal computable U with domain an antichain in $2^{<\omega}$, $K(\sigma)$ is the length of the shortest τ such that $U(\tau) = \sigma$. Similarly, for $X \in 2^\omega$, let $K^X(\sigma)$ denote the prefix-free Kolmogorov complexity of σ relative to X. That is, K^X is determined by a function universal among those computable relative to X.

Definition 1.8. A sequence $X \in 2^\omega$ is *K-trivial* if and only if there is a constant k such that for every ℓ, $K(X \upharpoonright \ell) \leq K(0^\ell) + k$, where 0^ℓ is the sequence of 0's of length ℓ.

By early results of Chaitin and Solovay and later results of Nies and others, there are a variety of equivalents to K-triviality and a variety of properties of the K-trivial sets. For example, X is K-trivial if and and only if for every sequence R, R is 1-random for λ if and only if R is 1-random for λ relative to X.

In the next section, we will apply the following.

Theorem 1.9 (Nies [Nie09], strengthening Chaitin [Cha76]). *If X is K-trivial, then there is a recursively enumerable and K-trivial set which computes X.*

The following lemma follows from the work of Nies and others [Nie09]. Some versions of this property have been used by Kučera extensively, e.g. in [Kuč85].

Lemma 1.10. *Suppose that X is K-trivial and $\{U_e^X : e \in \omega\}$ a uniformly $\Sigma_1^{0,X}$ family of sets. Then, there is a computable function g and a Σ_1^0 set V of measure less than 1 such for every e, if $\lambda(U_e^Z) < 2^{-g(e)}$ for every oracle Z, then $U_e^X \subseteq V$.*

Proof. Let $\left((E_i^e)\right)_{e\in\mathbb{N}}$ be a uniform sequence of all oracle Martin-Löf tests. A standard construction of a universal oracle Martin-Löf test (T_i) (e.g. see [Nie09]) gives a recursive function f such that $\forall Z \subseteq \omega\ (E_{f(i,e)}^{e,Z} \subseteq T_i^Z)$ for all $e, i \in \mathbb{N}$. Let $T := T_2$ and $f(e) := f(2, e)$ for all $e \in \mathbb{N}$, so that $\mu(T^Y) \leq 2^{-2}$ for all $Y \in 2^\omega$ and $E_{f(e)}^e \subseteq T$ for all $e \in \mathbb{N}$. In [KH07] it was shown that X is K-trivial iff for some member T of a universal oracle Martin-Löf test, there is a Σ_1^0 class V with $T^X \subseteq V$ and $\mu(V) < 1$.

Now given a uniform enumeration (U_e) of oracle Σ^0_1 classes we have the following property of T:

There is a recursive function g such that for each e,
either $\exists Z \subseteq \omega\ (\mu(U^Z_e) \geq 2^{-g(e)-1})$, or $\forall Z \subseteq \omega\ (U^Z_e \subseteq T^Z)$.

To see why this is true, note that every U_e can be effectively mapped to the oracle Martin-Löf test (M_i) where $M^Z_i = U^Z_e[s_i]$ and s_i is the largest stage such that $\mu(U^Z_e[s_i]) < 2^{-i-1}$ (which could be infinity). Effectively in e we can get an index n of (M_i). It follows that if $\mu(U^Z_e) < 2^{-f(n)-1}$ for all Z, then $U^X_e = M^X_{f(n)} = E^{n,X}_{f(n)} \subseteq T^X \subseteq V$. So $g(e) = f(n) + 1$ is as wanted. □

1.3. *Our results*

Intuitively, $X \in \mathrm{NCR}_1$ asserts that X is not effectively random relative to any continuous measure and X is K-trivial asserts that relativizing to X does not change the evaluation of randomness relative to the uniform distribution. In the next section, we connect the two notions.

Theorem 1.11. *Every K-trivial set belongs to* NCR_1.

A recursively enumerable (r.e.) set W is called *incomplete* if it does not compute the halting problem $\emptyset'$.

Theorem 1.12. *If W is an incomplete r.e. set and $X \leq_{\mathrm{T}} W$, then $X \in \mathrm{NCR}_1$.*

As we mentioned above, Theorem 1.12 implies Theorem 1.11, because every K-trivial set is computable from a r.e. K-trivial set, and every K-trivial set is incomplete. However we believe that the technique in the direct proof of Theorem 1.11 is of independent interest.

2. K-trivial sets and NCR_1

In this section we prove Theorem 1.11.

Let Y be K-trivial and let μ be a continuous measure with representation m; we want to show Y is not μ-random. Assume for a contradiction that Y is μ-random. By Theorem 1.9, let X be a recursively enumerable K-trivial sequence that computes Y. Let f be the iterated convergence function as defined above for the computable approximation to Y given by approximating X's computation of Y. Since X is recursively enumerable, X can compute the convergence function for its own enumeration and hence f is computable from X.

Let s_m be the granularity function for μ as represented by m. By Theorem 1.7, f eventually dominates s_m. By changing finitely many values of f, we may assume that f dominates s_m everywhere. So, we have that for every n

$$\mu([Y \upharpoonright f(n)]) \leq 2^{-n}.$$

Further, we may assume that f can be obtained as the limit of a computable function $f(n,s)$ such that for all s, $f(n-1,s) \leq f(n,s) \leq f(n,s+1)$.

We will build an m-test $\{S_i : i \in \omega\}$ which is Solovay-correct for μ and which Y does not pass, thereby concluding that Y is not μ-random. That is, we plan to build $\{S_i : i \in \omega\}$ to be a uniformly $\Sigma_1^{0,m}$ sequence of sets such that $\sum_{i \in \omega} \mu(S_i)$ is bounded and such that there are many co-finitely i for which $Y \in [S_i]$. Our construction will not be uniform.

X's K-triviality is exploited in the form of Lemma 1.10. Let V and g be given by Lemma 1.10 where $\{U_e^X : e \in \omega\}$ is a listing of all $\Sigma_1^{0,X}$ sets. We will build an oracle Σ_1^0 class U along the construction. We use the recursion theorem to assume that in advance we know an index e such that $U = U_e$. During the construction we will make sure that for every oracle Z, $\lambda(U^Z) < 2^{-g(e)}$. Lemma 1.10 then implies that $U^X \subseteq V$. Let a denote $g(e)$. Since the measure of V is less than 1, we may assume that a is large enough so that $\lambda(V) + 2^{-a} < 1$.

We use the approximation to X as a computably enumerable set to enumerate approximations to initial segments of Y into the sets S_i; we rely on the K-triviality of X to keep the total μ-measure of the S_i's bounded.

For each $n > a$ we have a requirement R_n whose task is to enumerate $Y \upharpoonright f(n)$ into S_n. Let $y_{n,s} = Y_s \upharpoonright f(n,s)$ the stage s approximation to $Y \upharpoonright f(n)$. Let $x_{n,s}$ be the initial segment of X_s necessary to compute $y_{n,s}$ and $f(n,s)$. So, if $y_{n,s+1} \neq y_{n,s}$, it is because $x_{n,s+1} \neq x_{n,s}$. In this case, $x_{n,s+1}$ is not only different than $x_{n,s}$, but also incomparable. At stage s, R_n would like to enumerate $y_{n,s}$ into S_n, but before doing that it will *ask for confirmation* using the fact that $U^X \subseteq V$. Since we are constrained to keep $\lambda(U^X)$ less than or equal to 2^{-a}, we will restrict R_n to enumerate at most 2^{-n} measure into U^X. The reason why we need a bit of security before enumerating a string in S_n is that we have to ensure that $\sum_i \mu(S_i)$ is bounded. For this purpose, we will only enumerate mass into S_n when we see an equivalent mass going into V.

Action of requirement R_n:

(1) The first time after R_n is initialized, R_n chooses a clopen subset of 2^ω, σ_n, of m-measure less than 2^{-n}, that is disjoint form V_s and $U_s^{X_s}$.

Note that since V and U^{X_s} have measure less than $\lambda(V)+2^{-a}<1$, we can always find such a clopen set. Furthermore we can chose σ_n to be different from the σ_i chosen by other requirements R_i, $i>a$. We note the value of σ_n might change if R_n is initialized.

(2) To *confirm* $x_{n,s}$, requirement R_n enumerates σ_n into $U^{x_{n,s}}$. Requirement R_n will not be allowed to enumerate anything else into U^{X_s} unless X_s changes below $x_{n,s}$. This way R_n is always responsible for at most 2^{-n} measure enumerated in U^{X_s}.

(3) Then, we wait until a stage $t>s$ such that
   (a) either $x_{n,s} \not\subseteq x_{n,t}$ (as strings),
   (b) or $\sigma_n \subseteq V_t$.

   Observe that if $x_{n,s}$ is actually an initial segment of X, then we will have $\sigma_n \subseteq U^X \subseteq V$. So, we will eventually find such a stage t.
   - In Case 3(a), we start over with R_n. Note that in this case σ_n has come out of U^{X_t}, and hence R_n is responsible for no measure inside U^{X_t} at stage t.
   - In Case 3(b), if $\mu([y_{n,t}])<2^{-n}$, enumerate $y_{n,t}$ into S_n. (Recall that we are allowed to use the representation of μ as an oracle when enumerating S_n.)

Since we only enumerate $y_{n,t}$ of μ-measure less than 2^{-n} when σ_n is enumerated in V, we have that

$$\sum_i \mu(S_i) \leq \lambda(V) < 1.$$

It is not hard to check that $\lambda(U^X) \leq \sum_{n=a+1}^{\infty} 2^{-n} = 2^{-a}$, so we actually have that $U^X \subseteq V$. Also notice that once $x_{n,s}$ is a initial segment of X, we will eventually enumerate σ_n into V and an initial segment of Y into S_n.

This completes the proof of Theorem 1.11.

3. Incomplete r.e. degrees and NCR$_1$

We turn to the proof of Theorem 1.12. Let W be an incomplete r.e. set, and let $X \leq_{\mathrm{T}} W$.

The fact that W is recursively enumerable and $X \leq_{\mathrm{T}} W$ implies that there is a recursive approximation $X=\lim_t X_t$ such that the modulus functions g_X is recursive in W, hence $f_X \leq_{\mathrm{T}} W$.

Suppose, for contradiction, that X is 1-random relative to a representation m of a continuous measure μ. By Theorem 1.7, by changing f_X at finitely many inputs, we obtain a function $f \leq_{\mathrm{T}} W$ which bounds the granularity function s_m. Let $h(n) = X \restriction f(n)$. So $h \leq_T W$, and for all n, $\mu([h(n)]) < 2^{-n}$.

Let J be a universal partial recursive function. For $n \in \mathbb{N}$, let $U_n = \{J(n)\}$ if $n \in \operatorname{dom} J$ and $J(n)$ is a binary string such that $\mu(J(n)) < 2^{-n}$. Otherwise, U_n is empty. Then the test U is recursively enumerable in m, and is correct for μ. Since X must pass U, we see that for all $n \in \operatorname{dom} J$, $h(n) \neq J(n)$.

The function h is diagonally nonrecursive. By Jockusch [Joc89], h computes a fixed-point-free function. This contradicts Arslanov's completeness criterion [Ars81], which states that an incomplete r.e. set cannot compute a fixed-point-free function.

This completes the proof of Theorem 1.12.

The question of which Δ^0_2 sets belong to NCR_1 remains open.

References

[Ars81] Marat Arslanov. On some generalizations of the xed-point theorem. *Soviet Mathematics*, 25:1-10, 1981. Translated from Izvestiya Vysshikh Uchebnykh Zavedeniĭ Matematika.

[Cha76] Gregory J. Chaitin. Information-theoretic characterizations of recursive infinite strings. *Theoret. Comput. Sci.*, 2(1):45–48, 1976.

[Joc89] Carl G. Jockusch, Jr. Degrees of functions with no fixed points. In *Logic, methodology and philosophy of science*, VIII (Moscow, 1987), Volume 126 of Stud. Logic Found. Math., pp. 191–201. North-Holland, Amsterdam, 1989.

[KH07] Bjørn Kjos-Hanssen. Low for random reals and positive-measure domination. *Proc. Amer. Math. Soc.*, 135(11):3703–3709 (electronic), 2007.

[Kuč85] Antonín Kučera. Measure, Π^0_1-classes and complete extensions of PA. In *Recursion theory week (Oberwolfach, 1984)*, volume 1141 of *Lecture Notes in Math.*, pages 245–259. Springer, Berlin, 1985.

[Mon05] Antonio Montalbán. *Beyond the Arithmetic.* PhD thesis, Cornell University, 2005.

[Nie09] André Nies. *Computability and randomness*, volume 51 of *Oxford Logic Guides.* Oxford University Press, Oxford, 2009.

[RS07] Jan Reimann and Theodore A. Slaman. Probability measures and effective randomness. preprint, 2007.

[RS08] Jan Reimann and Theodore A. Slaman. Measures and their random reals. preprint, 2008.

[RSte] Jan Reimann and Theodore A. Slaman. The structure of the never continuously random sequences. in preparation, no date.

LIMITWISE MONOTONIC FUNCTIONS AND THEIR APPLICATIONS

Rodney G. Downey

Department of Mathematics, Statistics, and Operations Research
Victoria University of Wellington, P.O. Box 600
Wellington 6140 New Zealand
rod.downey@ecs.vuw.ac.nz

Asher M. Kach

Department of Mathematics
University of Connecticut – Storrs
196 Auditorium Road, Storrs, CT 06269, USA
asher.kach@uconn.edu

Daniel Turetsky

Department of Mathematics, Statistics, and Operations Research
Victoria University of Wellington, P.O. Box 600
Wellington 6140 New Zealand
dan.turetsky@msor.vuw.ac.nz

We survey what is known about limitwise monotonic functions and sets and discuss their applications in effective algebra and computable model theory. Additionally, we characterize the computably enumerable degrees that are *totally limitwise monotonic*, show the support strictly increasing $\mathbf{0}'$-limitwise monotonic sets on $\mathbb{Q}$ do not capture the sets with computable strong η-representations, and study the limitwise monotonic spectra of a set.

Keywords and phrases: limitwise monotonic function.

1. Introduction

Early applications of computability theory for demonstrating that various processes in mathematics were algorithmically unsolvable tended to be

---

The first and second author's research was supported by The Marsden Fund of New Zealand, the latter via a Post-Doctoral Fellowship. The third author's research was supported by a VIGRE Fellowship and a Research Assistantship from the University of Wisconsin Graduate School.

rather crude codings of the halting problem into the relevant mathematical structure. A classical example is the Novikov-Boone proof of the undecidability of the word problem in finitely presented groups (see [3], [4], [5], [6], and [7] and [26]). In that proof, a finitely presented group is constructed around a given description of the quadruples of a Turing machine, as in the proof of the undecidability of the word problem for finitely presented semigroups (see [28]), and then algrebra is used to make the machines action faithfully represented in the group. As observed by Post (see [27]), in many contexts it is enough to simply code the halting problem as a set, such as the proof of Gödel's incompleteness theorem.

Computable model theory, and later reverse mathematics, pointed at encodings which were more complex. In computable model theory, traditionally we assume that we are given some structure whose elements are coded by the integers and whose open diagram is computable. In particular, a computable linear ordering would simply be $(A :\leq)$, where the universe $A \subseteq \omega$ is computable and the ordering $\leq$ is a computable relation on $A \times A$. Orderings are a natural arena to find codings other than the halting problem, since it is quite hard to code sets into them at low levels since little is definable with one quantifier. One of the first applications of more complex codings was due to Feiner (see [12]) who demonstrated how to code a Σ_3^0 set into a computable linear ordering via sizes of *finite maximal blocks*, i.e., a finite collection of points, all adjacent, such that the left and right endpoints are limit points. One of the classical applications of Feiner's Theorem is to construct a $\mathbf{0}'$-computable linear ordering not classically isomorphic to a computable one by applying this result in relativized form and noting that there are sets S which are $\Sigma_3^{\emptyset'}$-computable but not Σ_3^0-computable.

Later, Lerman studied such codings where the finite maximal blocks were separated by the order type ζ.

Definition 1.1. The *strong ζ-representation* of a set $S = \{n_0 < n_1 < n_2 < \dots\}$ is the linear order

$$\zeta + n_0 + \zeta + n_1 + \zeta + n_2 + \dots.$$

A *weak ζ-representation* of a set $S = \{n_0 < n_1 < n_2 < \dots\}$ is a linear order

$$\zeta + n_{f(0)} + \zeta + n_{f(1)} + \zeta + n_{f(2)} + \cdots$$

for some (total) surjective function f.

Theorem 1.2 (Lerman [25]). *A set S has a computable strong ζ-representation if and only if S is Δ^0_3. A set S has a computable weak ζ-representation if and only if S is Σ^0_3.*

There are many applications of the technique of coding higher level sets into algebraic invariants of some algebraic object. They include applications in (abelian) group theory, ring theory, logic, lattice theory, etc. Here we refer the reader to [2] for such examples.

Beginning with the work of Khisamiev (see [21]), more subtle considerations came into such codings when it was realized that *not only the arithmetical complexity of the set is important, but also the manner of the formation of the set.* Khisamiev's intuition was that computability is concerned with dynamic enumerations, and this fact has ramifications for objects in computable structures. Khisamiev's example was in abelian p-groups, and we will look at his example later, but the nature of our concern is best illustrated by asking which equivalence structures are computable. Note that if we have a computable equivalence relation $\equiv$, then $[x]_\equiv$, the equivalence class of x, has the following important property: *it only gets bigger!* That is, at the stage that x enters the universe of the relation, we might discover that $[x]_\equiv$ has n many elements. At later stages, the class can only gain elements. This phenomenon is captured by limitwise monotonicity.

Definition 1.3. A function F is *limitwise monotonic* if there is a computable approximation function $f(\cdot,\cdot)$ such that, for all x,

(i) $F(x) = \lim_s f(x,s)$.
(ii) For all s, $f(x,s) \leq f(x,s+1)$.

A set S is *limitwise montonic* if it is the range of a limitwise monotonic function.

It is a very easy argument to prove the following.

Theorem 1.4 (Calvert, Cenzer, Harizanov, and Morozov [8]). *An equivalence structure $\mathcal{E}$ with infinitely many classes is computable if and only if there is a limitwise monotonic function F (with range $\omega \cup \{\infty\}$) for which there are exactly $|\{x : F(x) = \kappa\}|$ many classes of size κ (for each $\kappa \in \omega \cup \{\infty\}$) in $\mathcal{E}$.*

We leave the proof of Theorem 1.4 to the reader. Theorem 1.4 can be rephrased to say *the computable isomorphism types of computable equivalence structures are specified by limitwise monotonic functions.*

It turns out that there are a number of applications of limitwise monotonicity in the literature, and we will explore some of these in this paper. They include quite a number of problems in linear orderings, trees, p-groups, computable spectra of $\aleph_1$-categorical structures, and general computable model theory in and around prime models. Many of these applications require that the notion be applied in relativized form, and we will explore this, proving some general theorems about such sets.

Because of the connection between limitwise monotonic sets and applications, we will try to understand when it is possible to find a nonlimitwise monotonic set below a given degree. We therefore introduce the following concept.

Definition 1.5. A degree $\mathbf{a}$ is *totally limitwise monotonic* if every set $B \leq_T \mathbf{a}$ is a limitwise monotonic set.

We prove the following result.

Theorem 1.6. *A computably enumerable degree* $\mathbf{a}$ *is totally limitwise monotonic if and only if* $\mathbf{a}$ *is non-high.*

Thus, for example, if a non-high computably enumerable degree $\mathbf{a}$ can compute a set, then there is a computable equivalence structure having classes of exactly those sizes.

Additionally, we will look at the situation where limitwise monotonicity seems not to be enough, but certain variations suffice, such as η-representations. Here we will show that for the question of strong η-representations, the variation of limitwise monotonicity introduced by Kach and Turetsky (see [20]) does not suffice.

Theorem 1.7 (Turetsky). *There is a set S with a computable strong η-representation that is not support strictly increasing $\mathbf{0}'$-limitwise monotonic on $\mathbb{Q}$.*

In the last section, we introduce a new notion which we term the *limitwise monotonic spectrum* of a set. The idea here is that we wish to recast the relationship between limitwise monotonicity and degrees of unsolvability in a more abstract setting. This leads to the following definition.

Definition 1.8. If $S \subseteq \omega$ is any nonempty set, define the*limitwise monotonic spectrum of S* (denoted $\mathrm{LMSpec}(S)$) to be the set

$$\mathrm{LMSpec}(S) := \{\mathbf{a} : S \text{ is } \mathbf{a}\text{-limitwise monotonic}\}.$$

In addition to rephrasing existing results, we show if $\mathbf{a} < \mathbf{b}$, then there is a set S with $\mathbf{a} \notin \mathrm{LMSpec}(S)$ and $\mathbf{b} \in \mathrm{LMSpec}(S)$. While we do not develop this subject further, we believe this may have wider applications.

2. Limitwise Monotonic Functions and Sets

As a first step towards understanding the limitwise monotonic sets, it is natural to determine where they sit in the arithmetic hierarchy. It is also important to recognize that being a limitwise monotonic set is not degree invariant.

Theorem 2.1 (Folklore).

(i) If A is a limitwise monotonic set, then A is Σ^0_2.
(ii) If A is Σ^0_2, then $A \oplus \omega$ is a limitwise monotonic set.

Proof. (i) Let f be a computable approximation function witnessing that A is a limitwise monotonic set. Then $n \in A$ if and only if $(\exists x)(\exists s)(\forall t \geq s)\,[f(x,t) = n]$.

(ii) Let A be Σ^0_2, so that $n \in A$ if and only if $(\exists s)(\forall t)\,[R(n,s,t)]$. For convenience, we assume $(\forall n)\,[\neg R(n,0,0)]$. Then

$$f(\langle n, s \rangle, t) = \begin{cases} 2n & \text{if } (\forall t' < t)\,[R(n,s,t')], \\ 2n+1 & \text{otherwise,} \end{cases}$$

witnesses that $A \oplus \omega$ is a limitwise monotonic set. □

The latter is essentially an idea of Lerman (see [25]). As implicitly noted there, this result generalizes to any nonimmune Σ^0_2 set in place of ω. Indeed, it is not hard to show any Σ^0_2 set containing an infinite limitwise monotonic set is itself a limitwise monotonic set.

It might be tempting to conjecture that every Σ^0_2 set is a limitwise monotonic set, but this is not true. Though every c.e. set is clearly limitwise monotonic, not every d.c.e. set is a limitwise monotonic set. Recall a set A is *d.c.e.* if there are c.e. sets B and C with $A = B - C$.

Theorem 2.2 (Khoussainov, Nies, and Shore [23]). *There is a Δ^0_2 set A, indeed a d.c.e. set A, which is not a limitwise monotonic set.*

Proof (Sketch). We construct a d.c.e. set A satisfying the requirements:

$\mathcal{R}_e$: The function $\varphi_e(\cdot,\cdot)$ does not witness that A is a limitwise monotonic set.

Towards meeting $\mathcal{R}_0$, we pick a witness n_0 and put n_0 into A. We then wait for $\varphi_0(x, t_0) = n_0$ for some x and stage $t_0 > n_0$. At such a stage t_0, we remove all elements between n_0 and t_0 currently in A, and allow only elements n larger than t_0 to enter A. We maintain this $\mathcal{R}_0$ restraint until a stage $t_1 > t_0$ and number $n_1 > n_0$ is found with $\varphi_0(x, t_1) = n_1$ and $n_1 \in A$. At such a stage t_1, we put $n_0 + 1$ into A. We then repeat this process by taking all elements between n_1 and t_1 out of A, releasing the previous A restraint, and allowing only elements n less than t_0 or bigger than t_1 to enter A.

If for every i we find a stage t_i and number n_i, then the values of $\varphi_0(x, s)$ tend to infinity, satisfying $\mathcal{R}_0$. Lower priority strategies $\mathcal{R}_e$ can then play "behind" this $\mathcal{R}_0$ restraint. If instead there is an i such that t_i and n_i are never found, then $\lim_s \varphi_0(x, s) \notin A$, satisfying $\mathcal{R}_0$. Lower priority strategies can then play "above" this finitary restraint.

The interaction of strategies is straightforward, with each strategy guessing whether higher priority strategies have a finitary or infinitary outcome. □

Thus, limitwise monotonicity is not degree invariant as the set A of Theorem 2.2 is not limitwise monotonic, yet $A \oplus \omega$ is limitwise monotonic by Theorem 2.1.

In the definition of a limitwise monotonic set, we imposed no restraint on the number of times an element could appear in the range of F. One might think that such a restraint would give a stronger notion, but this turns out not to be the case.

Theorem 2.3 (Harris [14]). *If F is a limitwise monotonic function with infinite range, then there is an injective limitwise monotonic function G with range(F) = range(G).*

Proof. Fixing a limitwise monotonic approximation f for F, we define a limitwise monotonic approximation g for G. Indeed, we define

$$g(m,s) = \begin{cases} 0 & \text{if } m \geq s, \\ f(n,t) & \text{otherwise, where } \langle n,t\rangle \text{ is least with } t \geq s \text{ so that} \\ & \quad f(n,t) \neq g(x,s) \text{ for all } x < m \text{ and } f(n,t) \geq g(m,s-1). \end{cases}$$

As g is clearly computable (note $\langle n, t\rangle$ must exist as range(F) is infinite) and nondecreasing, it suffices to argue that $G(m) = \lim_s g(m, s)$ exists for all m, that G is injective, and that range(F) = range(G).

Induction demonstrates the limit $G(m)$ exists for all m. Indeed, if $G(0)$, $\ldots$, $G(m-1)$ all exist, fix a stage after which these limits are achieved. Then either $g(m, s) < \max\{G(0), \ldots, G(m-1)\}$ for all s (in which case $G(m)$ exists) or there exists an $s_0 \in \omega$ with $g(m, s_0) > \max\{G(0), \ldots, G(m-1)\}$. In the latter case, the value of n used in defining $g(m, s)$ can only change finitely often, from which it follows that $G(m)$ exists.

The injectivity of G follows from g being injective at every stage, i.e., that $g(m, s) \neq g(m', s)$ if $m \neq m'$.

Finally, we argue range(F) = range(G). As range$(G) \subseteq$ range(F) is immediate, we demonstrate that $F(n) \in$ range(G) for all $n \in \omega$ by induction on n. If $F(0), \ldots, F(n-1) \in$ range(G), fix a stage s_0 for which $F(k) = f(k, s)$ for all $s \geq s_0$ and $k \leq n$ and for which $g(m, s) = F(k)$ for all $s \geq s_0$ and m witnessing $F(0), \ldots, F(n-1) \in$ range(G). Then at stage s_0 (provided $s_0 > n$), either $g(x, s_0) = f(n, s_0)$ for some $x < s_0$ or $g(s_0, s_0)$ will be defined as $F(n) = f(n, s_0)$. In the former case, either $G(x) = F(n)$ or $G(y) = F(n)$ for some $y < x$; in the latter case, either $G(s_0) = F(n)$ or $G(y) = F(n)$ for some $y < s_0$. Thus $F(n) \in$ range(G). □

Because of its use in applications, it is natural to wonder which degrees compute a set that is not a limitwise monotonic set. This motivates our notion of a *totally limitwise monotonic degree* (recall Definition 1.5).

Theorem 2.4. *A computably enumerable degree* **a** *is totally limitwise monotonic if and only if* **a** *is non-high.*

Proof ($\Longrightarrow$). We show if **a** is high, then it computes a nonlimitwise monotonic set. Fixing a c.e. set $A \in$ **a** with enumeration $\{A_s\}_{s \in \omega}$, as **a** is high there is an A-computable function f^A that is dominating with respect to the class of all computable functions. Denote the use of f^A by g, so that $g(n)$ is the use of the computation $f^A(n)$. We build a set $S \leq_T A$ (we witness this by $S = \Gamma^A$) that is not a limitwise monotonic set by diagonalizing against all candidate approximation functions $\{\varphi_i\}_{i \in \omega}$. We introduce the obvious requirements.

$\mathcal{R}_i$: The function Φ_i does not witness that S is a limitwise monotonic set.

The strategy to satisfy $\mathcal{R}_i$ is similar to that in Theorem 2.2 except here we need an A-permission to remove a witness n_0 put into S. As this A-permission may never appear, there is a need to choose another witness n_1 and restart this process. We build a computable function h, extending it whenever we need a permission. That f^A is dominating will guarantee that eventually a witness will receive an A-permission, else it would be the case that f^A would not dominate h.

Strategy for $\mathcal{R}_i$:

(1) Let $k := 0$.
(2) Let s_k be the stage at which the value of k being used was assigned.
(3) Choose an integer n_k and put n_k in B by defining $\Gamma^{A \upharpoonright g(n_k)}(n_k)[s_k] := 1$.
(4) Wait for a column x_k for which $\varphi_i(x_k, s) = n_k$.
(5) For $n \leq n_k$, if $h(n)$ is not yet defined, define $h(n) := f^{A_s}(n_k)$.
(6) Wait for $A_s \upharpoonright (g(n_k)[s_k]) \neq A_{s_k} \upharpoonright (g(n_k)[s_k])$. While waiting, increment k and return to Step 2.
(7) If an A-permission is seen, remove n_ℓ from B for all $\ell \geq k$ by defining $\Gamma^{A_{s_k} \upharpoonright g(n_k)}(n_\ell) := 0$, and cancel all work for all n_ℓ with $\ell > k$.
(8) Keep n_ℓ out of B for all $\ell \geq k$, and wait for $\varphi_i(x_k, s)$ to increase to some m currently in B.
(9) Return to Step 5 with m in place of n_k and s in place of s_k.

Of course, if the strategy spends cofinitely many stages in Step 4 with some n_k, then $\mathcal{R}_i$ is satisfied as $n_k \in S$ and $n_k \notin \text{range}(\Phi_i)$. If the strategy spends cofinitely many stages in Step 8 for some n_k, then $\Phi_i(x_k) \notin S$. If the strategy returns to Step 5 infinitely many times for some k, then $\mathcal{R}_i$ is satisfied as $\Phi_i(x_k)$ is not defined (having an infinite limit). However, one of these must be the case, as if we wait at Step 6 for cofinitely many stages for all k, then h is not dominated by f^A, a contradiction.

As before, the interaction of strategies is straightforward, with each strategy guessing whether higher priority strategies have a finitary or infinitary outcome. As there are (infinitely many) functionals that are empty on any oracle, the resulting set B is nonempty (indeed infinite). □

Proof ($\Longleftarrow$). Fixing a nonhigh c.e. degree $\mathbf{a}$ and a c.e. set $A \in \mathbf{a}$ with approximation $\{A_s\}_{s\in\omega}$, we show every infinite $B \leq_T A$ (say $B = \Psi^A$) is a limitwise monotonic set. We approximate B by running Ψ^A with the approximations to A. Let B_s denote our approximation at stage s, i.e., let $B_s = \Psi^A[s]$. We enumerate a subset of B_s recursively as follows:

$$b_0^s = \min B_s$$
$$b_{n+1}^s = \min\{b \in B_s \mid (\forall t \leq s)[b > b_n^t]\}.$$

Before continuing, we argue $\lim_s b_n^s$ exists for all n.

Claim 2.4.1. For every n, the sequence $\{b_n^s\}_{s\in\omega}$ converges to a finite limit.

Proof. Clearly, we have $\min B = \lim_s b_0^s$. Assuming $\lim_s b_n^s$ exists, we show $\lim_s b_{n+1}^s$ exists. As $\lim_s b_n^s$ exists by hypothesis, the set $\{b_n^s\}_{s\in\omega}$ is finite, so we may let $b \in B$ be least such that b is greater than all these values. Let s_0 be a stage by which B has converged on b and b_n^s has converged. Then for any stage $s > s_0$, we have $b_{n+1}^s = b$. □

With this, we define a total function $f = \Gamma^A$. Using a witness to the failure of f being a dominating function for the class of total computable functions, we demonstrate B is a limitwise monotonic set. As preparation, let $\{\varphi_i\}_{i\in\omega}$ be an effective listing of all partial computable functions with the property that $\varphi_{i,s}(n)\downarrow$ implies $\varphi_{i,s}(m)\downarrow$ for all $m < n$.

Initially, we define $\Gamma^{A_0}(n) = 0$ for all n with use $\psi(b_n^0)$. At stage $s+1$, if $\Gamma^{A_s}(n)$ is undefined, we define it by

$$\Gamma^{A_s}(n) = \max\{\varphi_{i,s}(n) : i < n \text{ and } \varphi_{i,s}(n)\downarrow\} + 1$$

with use $\psi(b_n^s)$.

As a consequence of Claim 2.4.1, it follows that $\psi(b_n^s)$ converges. Thus f is a total A-computable function. As **a** is nonhigh, we may fix an index k such that φ_k is a total computable function which f does not dominate. From this, as already suggested, we construct a limitwise monotonic approximation h to B.

Construction: At stage $s = 0$, we define $h(x, 0) = 0$ for all x.

At stage $s+1$, if φ_k has not converged on any new values since stage s, we define $h(x, s+1) = h(x, s)$ for all x. If $\varphi_k(n)$ newly converges at stage s for some $n \leq s$, then for every $b \in B_s$ with $b \leq b_n^s$, we choose a previously unused x and define $h(x, s+1) = b$. For every x with $h(x, s) \notin B_s$, we define $h(x, s+1) = b_m^s$ for the least $m \leq n$ with $b_m^s > h(x, s)$. For all other $x \leq s$, we define $h(x, s+1) = h(x, s)$.

Verification: We verify that h is a limitwise monotonic approximation to B. By construction, the function h is total and computable. Define $H(x) = \lim_s h(x, s)$. We verify that $H(x)$ exists and that H witnesses that B is a limitwise monotonic set.

Claim 2.4.2. The function H is total.

Proof. Fixing an integer x, we may suppose there is a stage s_0 with $h(x, s_0) > 0$ (else $H(x) = 0$). Let n be an integer such that $\varphi_k(n)$ has not converged by stage s_0, and $\varphi_k(n) \geq f(n)$. Let $s_1 > s_0$ be the stage at which $\varphi_k(n)$ converges. Then b_n^s has necessarily converged by stage s_1. By our assumption that $\varphi_k(n)$ converges before $\varphi_k(m)$ for any $m > n$, we have $h(x, s_1) \leq b_n^{s_1}$. Thus $H(x) \leq b_n^{s_1}$. □

Claim 2.4.3. The function H enumerates B.

Proof. For any $b \in B$, choose a stage s_0 such that B_s has converged on b, and an n such that $b_n^{s_0} > b$. When φ_k converges on any $m > n$, an x will be created such that $h(x, s) = b$, and this will never change at later s. Thus $H(x) = b$.

For any $c \notin B$, choose a stage s_0 such that B_s has converged on c, and an n such that $b_n^{s_0} > c$. When φ_k converges on any $m > n$, any x with $h(x, s) = c$ will change their value. Thus, for no x does $H(x) = c$. □

This completes the proof.

The hypothesis that **a** is c.e. in Theorem 2.4 is very much necessary as the following result shows.

Theorem 2.5 (Hirschfeldt, R. Miller, and Podzorov [15]). *There is a low Δ_2^0 set A which is not a limitwise monotonic set.*

Proof (Sketch). In addition to the $\mathcal{R}_e$ requirements of Theorem 2.2, we meet the standard lowness requirements:

$$\mathcal{N}_e : \exists^\infty s \left[\Psi_e^A(e)[s]\downarrow\right] \implies \Psi_e^A(e)\downarrow .$$

To allow $\mathcal{N}_e$ to be met in the presence of the $\mathcal{R}_e$ requirements of Theorem 2.2, we use the fact that A is Δ_2^0. More specifically, we allow $\mathcal{R}_j$ for $j < e$ to injure $\mathcal{N}_e$. This would seem problematical since the action of $\mathcal{R}_j$ may be infinitary. However, this is not the case.

For example, consider a single higher priority $\mathcal{R}_j$ requirement. If we are at a stage where we see some computation $\Psi_e^A(e)[s]\downarrow$ and the use $\psi_e^A(e)[s]$ is less than the number n_i currently being used for $\mathcal{R}_j$, then $\mathcal{N}_e$ can assert control of $A \restriction \psi_e^A(e)[s]$ and preserve the computation with impunity. On the other hand, it might be that $\mathcal{R}_j$ is pointing at some n_i below the use $\psi_e^A(e)[s]$. In this case, what $\mathcal{N}_e$ will do is assert control and restrain this portion of A with priority e. Of course $\mathcal{R}_j$ can later injure this, but when it does it must move to a new n_{i+1} and this will be large. It might happen that at some stage $s' > s$, again $\mathcal{N}_e$ might try to preserve some new computation $\Psi_e^A(e)[s']\downarrow$ and this new computationcan be injured by $\mathcal{R}_j$

again. *But* once we pick yet another n_{i+2}, we note that $\mathcal{R}_j$ will only be concerned with numbers *bigger* than $n_{i+1} > \psi_e^A(e)[s]$. With no injury to $\mathcal{R}_j$ we can therefore *restore* the stage $A \restriction (\psi_e^A(e)[s])[s'] = A \restriction \psi_e^A(e)[s]$ and then this computation $\Psi^A(e)[s'] = \Psi^A(e)[s]$ with no possible future injury from $\mathcal{R}_j$.

In this way, with a finite injury argument we can meet all the $\mathcal{N}_e$. □

The final result in this section is a basis theorem for Π_1^0 trees. Its proof is not illuminating, so we omit it.

Theorem 2.6 (J. Miller [20]). *If $P \subseteq 2^\omega$ is a Π_1^0 class containing a nonempty set, then P contains a limitwise monotonic set.*

Thus, for example, there are limitwise montonic sets that are 1-random and that are DNR_2.

3. Applications of Limitwise Monotonic Functions and Sets

Though equivalence structures appear to be the simplest application of limitwise monotonic sets to effective algebra, historically limitwise monotonic functions were introduced by Khisamiev as a means of characterizing the computable reduced abelian p-groups of length ω (see [21]). There, these functions were termed *s-functions.*

Theorem 3.1 (Khisamiev [21]). *A reduced abelian p-group $\mathcal{G}$ of length ω is computable if and only if there is a limitwise monotonic function F such that $u_n(\mathcal{G}) = |\{x : F(x) = n\}|$.*

Though the proof of Theorem 3.1 is more involved than Theorem 1.4, the idea is very much the same. The height of an element x in the group $\mathcal{G}[s]$ is computable from a computable presentation, and this height can only increase as s increases. Conversely, it is easy to build $\mathcal{G}$ from a computable approximation f to the limitwise monotonic function F.

After Khisamiev's work, other applications were discovered in effective algebra, particularly within the context of linear orders. The simplest (infinite) orderings are presentations of ω. The simplest relation on such orderings would be a unary relation. This began with Downey, Khoussainov, J. Miller, and Yu (see [11]) (which was circulated for a long time in preprint form), then Hirschfeldt, R. Miller, and Podzorov (see [15]), and finally Knoll (see [24]).

Definition 3.2 (Hirschfeldt, R. Miller, and Podzorov [15]). *A set A is* order computable *if there is a computable copy of $(\omega : <, A)$ in the language of linear orderings with an additional unary predicate.*

Observation 3.3 (Kach and Turetsky [20]). *Every order computable set is a limitwise monotonic set.*

The proof of this observation is immediate. When we see some element n declared to be in the set representing A in the computable copy of $(\omega : <, A)$, then n can only move to bigger things and has a limit. Since every c.e. set is limitwise monotonic, the following result says that order computability is a significantly more refined concept than limitwise monotonicity.

Theorem 3.4 (Downey, Khoussainov, J. Miller, and Yu [11]). *Every high c.e. degree contains a c.e. set which is not order computable.*

The last applications we give of (unrelativized) limitwise monotonicity are within computable model theory. Baldwin and Lachlan showed that for an uncountably categorical but not countably categorical theory T, the countable models form an elementary chain of length $\omega + 1$. An interesting line of research has been determining whether, for a fixed set $S \subseteq \omega + 1$, there is a theory T whose computable models (identifying a model with its position in the elementary chain) are precisely those in S. Khoussainov, Nies, and Shore realized the subset $(\omega + 1) - \{0\}$ via limitwise monotonic sets. Hirschfeldt, Khoussainov, and Semukhin realized the set $\{\omega\}$ via a variant of limitwise monotonic sets.

Theorem 3.5 (Khoussainov, Nies, and Shore [23]). *There is an uncountably categorical but not countably categorical theory T for which every model but the prime model is computable (realizing the set $(\omega + 1) - \{0\}$).*

Proof (Sketch). The desired theory T is in the language of infinitely many binary predicates $\{P_i\}_{i \in \omega}$. For each $n \in \omega$, an *n-cube* is a collection of 2^n many elements isomorphic to the structure with universe $\{x_\tau\}_{\tau \in 2^n}$ satisfying $P_i(x_\tau, x_{\tau'})$ if and only if $x_\tau(i) \neq x_{\tau'}(i)$ and $x_\tau(j) = x_{\tau'}(j)$ for all $j < i$. An *ω-cube* is the union of a chain of n-cubes for all $n \in \omega$.

From a computable structure consisting only of finite n-cubes, it is possible to approximate from below the maximal integer n to which a given element is part of an n-cube. Thus a set $S \subseteq \omega$ is a limitwise monotonic set if the structure consisting of (exactly) one n-cube for each $n \in S$ is computable.

Conversely, given an approximation of an integer n from below, it is possible to uniformly construct a computable presentation of an n-cube. Thus if S is a limitwise monotonic set, the structure consisting of (exactly) one n-cube for each $n \in S$ is computable.

It therefore suffices to fix a Σ_2^0 set S that is not a limitwise monotonic set. Let T be the theory of the model containing (exactly) one n-cube for each $n \in S$. Any non-prime model contains an ω-cube, and is thus computable as any wrongly constructed n-cubes can be grown into a fixed ω-cube. On the other hand, the prime model cannot be computable as S was chosen not limitwise monotonic. □

Definition 3.6. (Hirschfeldt, Khoussainov, and Semukhin [17]) Let $S \subseteq \omega$ be infinite. An *S-limitwise monotonic function* is a function $F : \omega \to [\omega]^{<\omega} \cup \{\infty\}$ for which there is a computable function $f : \omega \times \omega \to [\omega]^{<\omega} \cup \{\infty\}$ such that, for all n,

(i) $F(n) = \lim_s f(n, s)$.
(ii) For all s, if $f(n, s+1) \neq \infty$, then $f(n, s) \subseteq f(n, s+1)$.
(iii) For all s, if $f(n, s) = \infty$, then $f(n, s+1) = \infty$.
(iv) For all s, if $f(n, s) \neq \infty$ and $f(n, s+1) = \infty$, then $f(n, s) \subset S$.

A collection of finite sets $\{A_i\}_{i\in\omega}$ is *S-limitwise monotonic* if $\{A_i\}_{i\in\omega} = \{F(n) : F(n) \neq \infty\}$ for some S-limitwise monotonic function F.

Lemma 3.7 (Hirschfeldt, Khoussainov, and Semukhin [17]). *There is an infinite computably enumerable set S and uniformly computably enumerable sets $\{A_i\}_{i\in\omega}$ such that*

(i) Each A_i is either finite or is S.
(ii) If $x \in S$, then $x \in A_i$ for almost all i.
(iii) If $x \notin S$, then $x \notin A_i$ for almost all i.
(iv) If A_i is finite, then there is a $k \in A_i$ with $k \notin A_j$ for $j \neq i$.
(v) The collection $\{A_i : |A_i| < \aleph_0\}$ is not S-limitwise monotonic.

Theorem 3.8 (Hirschfeldt, Khoussainov, and Semukhin [17]).
There is an uncountably categorical but not countably categorical theory T for which only the saturated model is computable (realizing the set $\{\omega\}$).

Proof (Sketch). The desired theory T is in the language of infinitely many binary predicates $\{P_i\}_{i\in\omega}$. For each $n \in \omega$, an (n)-cube is the two element structure such that $P_i(x, y)$ if and only if $i = n$ and $x \neq y$. If $n_1, \ldots, n_k, n_{k+1}$ are all distinct, a $(n_1, \ldots, n_k, n_{k+1})$-cube is the union of two disjoint $(n_1, \ldots, n_k)$-cubes (defined by induction) with the additional relations $P_{n_{k+1}}(x, y)$ if $y = \pi(x)$, where π is some fixed bijection between the two $(n_1, \ldots, n_k)$-cubes.

Fix a set S and sequence of sets $\{A_i\}_{i\in\omega}$ as in Lemma 3.7. In a manner similar to Theorem 3.5, let T be the theory of the model containing an A_i-

cube for each $i \in \omega$ with $|A_i| < \aleph_0$. Then, as the sets A_i are uniformly computably enumerable, it is straightforward to see that the saturated model of T is computable. Also, as $\{A_i : |A_i| < \aleph_0\}$ is not S-limitwise monotonic, no nonsaturated model of T can be computable. □

4. Relativized Limitwise Monotonicity

For many other applications of limitwise monotonicity in effective algebra and computable model theory, the notion needs to be relativized. In many of these cases, a relativized version of Theorem 2.2 is useful.

Definition 4.1. A function F is $\mathbf{a}$-*limitwise monotonic* if there is an $\mathbf{a}$-computable approximation function $f(\cdot,\cdot)$ such that, for all x,

(i) $F(x) = \lim_s f(x,s)$.
(ii) For all s, $f(x,s) \leq f(x,s+1)$.

A set S is $\mathbf{a}$-*limitwise montonic* if it is the range of a $\mathbf{a}$-limitwise monotonic function.

Corollary 4.2. *There is a Δ^0_3 set A which is not $\mathbf{0}'$-limitwise monotonic.*

Perhaps the most direct application of $\mathbf{0}'$-limitwise motonicity is to linear orderings. In the same spirit as ζ-representations (see Definition 1.1), we can define the *strong η-representation* of an infinite set S and the *weak η-representation* by replacing the order type ζ with the order type η. It is easy to see that the set of maximal block sizes is a $\mathbf{0}'$-limitwise monotonic set if the η-representation is computable, as $\mathbf{0}'$ can decide if a pair of points form an adjacency. Partially answering a question of Downey (see [10]), Harris established the following.

Theorem 4.3 (Harris [14]). *A set S has a computable weak η-representation if and only if S is $\mathbf{0}'$-limitwise monotonic.*

We will sketch the proof of Theorem 4.3. We begin with a result independently proven by Harris and Kach.

Proposition 4.4 (Harris [14] and Kach [18]). *A function F is $\mathbf{0}'$-limitwise monotonic if and only if there is a computable function $g(\cdot,\cdot)$ such that $F(n) = \liminf_s g(n,s)$.*

Kach defined a set A to be a *limit infimum set* if there is a computable function g as in Proposition 4.4 with A the range of $\liminf_s g(\cdot,s)$. Then Proposition 4.4 can be rephrased as *a set A is a limit infimum set if and only if A is a $\mathbf{0}'$-limitwise monotonic set.*

Proof of Proposition 4.4. Let f witness that F is $\mathbf{0}'$-limitwise monotonic. By the Limit Lemma, there is a computable function h such that $f(n,s) = \lim_t h(n,s,t)$ for all n and s. Fixing n, the idea is to view $h(n,0,t)$, $h(n,1,t)$, ..., $h(n,t,t)$ as approximations to $F(n)$. We define $g(n,t)$ to be the maximum value of $h(n,j,t)$ for $j < i$, where i is maximal so that $h(n,i,t) = h(n,i,t-1)$. It can be verified that g is indeed a limit infimum approximation to F with the property $F(n) = G(n) := \liminf_t g(n,t)$.

For the reverse direction, let g witness that F is limit infimum. Defining $f(n,s)$ by $f(n,s) = \min\{g(n,t) : t \geq s\}$ yields an approximation function that is readily verified to be a $\mathbf{0}'$-limitwise monotonic approximation to F with the property $F(n) = \lim_s f(n,s)$ where $F(n) := \liminf_t g(n,t)$. □

Proof of Theorem 4.3. Via Proposition 4.4, we may fix a function g witnessing that S is limit infimum. Then, for each n, we will build a block of size $\liminf_s g(n,s)$ in stages, putting a dense ordering between the blocks. The action at stage $s+1$ depends on the relative sizes of $g(n,s)$ and $g(n,s+1)$: if $g(n,s+1) > g(n,s)$, we can add to the *outside* of the current $g(n,s)$ block so that it has size $g(n,s+1)$; and if $g(n,s+1) < g(n,s)$, we can remove the *outside* points and incorporate them into the adjacent interval of order type η. Independent of the relative sizes, we work towards making the endpoints limit points with the order type η between adjacent blocks. □

Analyzing the proof a bit more carefully, it is easy to see that the maximal blocks created appear in the same order (and thus with the same multiplicity) as given by F. Using Theorem 2.3, Harris was thus able to characterize the sets S with a computable *unique η-representation*, i.e., any weak η-representation where the function f in Definition 1.1 (with the order type η replacing the order type ζ) is injective.

Corollary 4.5 (Harris [14]). *A set S has a computable unique η-representation if and only if S is $\mathbf{0}'$-limitwise monotonic.*

Another application of $\mathbf{0}'$-limitwise monotonicity in linear orderings concerns shuffle sums. Recall that the *shuffle sum* of a set S is a linear ordering obtained by taking the rationals and replacing each element by a block of cardinality a member of S in such a way that the blocks representing the members of S occur densely.

Theorem 4.6 (Kach [18]). *The shuffle sum of a set S is computable if and only if S is $\mathbf{0}'$-limitwise monotonic.*

Proof (Sketch). Via Proposition 4.4, again fix a function g so that S is the range of $\liminf_s g(\cdot, s)$. This time we build blocks of size $g(n, s)$ densely at every stage. Again, we increase or cut back the size of blocks depending on whether $g(n, s)$ increases or decreases. The only complication is points which are removed cannot be so easily incorporated as in Theorem 4.3. Instead, new blocks will recycle these rejected points based on a priority ranking. □

Part of the interest for understanding which shuffle sums are computable stems from an earlier result in computable model theory by Hirschfeldt, answering a question of Rosenstein (see [30]).

Theorem 4.7 (Hirschfeldt [16]). *There is a complete theory T in the language of linear orders having a prime model and a computable model, but no computable prime model.*

Proof. By Corollary 4.2, fix a set $S \in \Sigma^0_3$ that is not $\mathbf{0}'$-limitwise monotonic. Let T be the theory of the shuffle sum of the set S. Then the shuffle sum of S is the prime model; however it is not computable as S was not $\mathbf{0}'$-limitwise monotonic. On the other hand, the shuffle sum of S with ζ is computable and is a model of T. □

The earliest application of $\mathbf{0}'$-limitwise monotonicity to linear orderings seems to concern isomorphism types of initial segments (i.e., convex initial sets) of linear orderings. Initial segments of linear orderings can be extremely complex, as witnessed, for example, by the initial segment of order type ω_1^{CK} of the Harrison ordering $\omega_1^{\mathrm{CK}}(1+\eta)$. Of interest here is how complex the classical isomorphism type of an initial segment of a computable ordering can be to still be assured a computable presentation. For example, Rosenstein asked whether every Π^0_2 initial segment of a computable linear ordering is isomorphic to a computable linear ordering (see [30]). Rosenstein had already demonstrated that this could not be strengthened to Π^0_3 using index sets (like Feiner's Theorem, see [12]). From the other direction, Raw showed every Σ^0_1 initial segment of a computable linear ordering is isomorphic to a computable linear ordering (see [29]). This was improved by Ambos-Spies, Cooper, and Lempp to every Σ^0_2 initial segment of a computable linear ordering has a computable copy (see [1]). Coles, Downey, and Khoussainov closed the gap, answering Rosenstein's question, by exhibiting a computable linear ordering with a Π^0_2 initial segment not isomorphic to a computable linear ordering. The initial segment is an *η-like* linear order, i.e., the result of replacing every element of the rationals with a finite block.

Theorem 4.8 (Coles, Downey, and Khoussainov [9]). *A set $S \subseteq \omega$ is Σ_3^0 if and only if there is a computable linear ordering $\mathcal{L}$ of the form $\mathcal{L} = \mathcal{A}+\mathcal{B}$ with $\mathcal{A}$ an η-like linear order having maximal blocks of sizes exactly those numbers in S and $\mathcal{B} \cong \omega^*$.*

The proof of Theorem 4.8 requires a reasonably nontrivial Π_2^0 priority argument (see [9] for details). But, granted Theorem 4.8, we can deduce the following.

Theorem 4.9 (Coles, Downey and Khoussainov [9]). *There is a computable linear ordering with a Π_2^0 initial segment not isomorphic to a computable linear ordering.*

Proof. By Corollary 4.2, there is a Σ_3^0 set S which is not a $\mathbf{0}'$-limitwise monotonic set. By Theorem 4.8, there is a computable linear ordering $\mathcal{L} = \mathcal{A} + \mathcal{B}$. Then $\mathcal{A}$ is not computable (as an isomorphism type) as S was not $\mathbf{0}'$-limitwise monotonic. However, the set A is Π_2^0, being the set of points having infinitely many points to the right. □

The last application within the context of linear orderings involves the complexity of subsets rather than the complexity of intervals. Kach and J. Miller used relativizations of limitwise monotonic functions to each degree $\mathbf{0}^{(n)}$ for $n \in \omega$ to prove the following result.

Theorem 4.10 (Kach and J. Miller [19]). *There is a computable non-well-ordered intrinsically computably well-ordered linear order, i.e., there is a computable non-well-ordered linear order for which no computable presentation has a computable subset of order type ω^*.*

Though the proof of Theorem 4.10 is rather involved, the major idea is that $\mathbf{0}^{(2n+1)}$ can approximate the value of $F(n)$ in a linear ordering of the form $\cdots + \omega^n \cdot F(n) + \cdots + \omega^2 \cdot F(2) + \omega \cdot F(1) + F(0)$ in a monotonic manner.

Despite all the discussed applications of relativized limitwise monotonic functions being in the context of linear orderings, Khisamiev first relativized limitwise monotonic functions in the context of reduced abelian p-groups.

Theorem 4.11 (Khisamiev [22]). *A reduced abelian p-group $\mathcal{G}$ of length less than ω^2 (say of length at most $\omega \cdot N$) is computable if and only if there are functions $F_0, F_1, \ldots, F_{N-1}$ such that F_i is $\mathbf{0}^{(2i)}$-limitwise monotonic and $u_{\omega \cdot i+n}(\mathcal{G}) = |\{x : F_i(x) = n\}|$.*

Proof (Sketch). If $\mathcal{G}$ has length less than ω^2, (nonuniformly) fix elements $g_1, g_2, \ldots, g_N$ with the height of g_i being $\omega \cdot i$. The oracle $\mathbf{0}^{(2i)}$ is powerful enough to approximate the height of elements below g_i, yielding a limitwise monotonic approximation function. Appealing to the remarks after Theorem 2.1, it follows (uniformly, though such uniformity is unnecessary) there is such a sequence of functions F_i. $\square$

5. Beyond Limitwise Monotonicity

As with order computable sets, sometimes (relativized) limitwise monotonicty fails to fully capture some algebraic phenomenon. A recent example of this was demonstrated by Kach and Turetsky in their work on an old question going back to Rosenstein (see [30]) and Downey (see [10]). Generalizing the notion of a strong η-representation, an *increasing η-representation* of S is a linear order $\mathcal{L}$ of the form $\eta + n_0 + \eta + n_1 + \dots$ where the n_i enumerate S in increasing order (possibly with repeats). It is not hard to show that not all $\mathbf{0}'$-limitwise monotonic sets S have increasing η-representations since such an S needs to be Δ^0_3.

To analyze the question of what $\mathbf{0}'$-limitwise monotonic sets S have strong or increasing η-representations, we introduce a new class of sets.

Definition 5.1 (Kach and Turetsky [20]). *A function $F : \mathbb{Q} \to \omega$ is* support (strictly) increasing *if $F(q_1) \leq F(q_2)$ ($F(q_1) < F(q_2)$) whenever $q_1 < q_2$ and $F(q_1), F(q_2) > 0$, the range of F is unbounded, and the support of F has order type ω.*

A function $F : \mathbb{Q} \to \omega$ is support (strictly) increasing limitwise monotonic on $\mathbb{Q}$ *if it is support (strictly) increasing and there is a computable approximation function $f : \mathbb{Q} \times \omega \to \omega$ such that $F(q) = \lim_s f(q, s)$ and $f(q, s) \leq f(q, s + 1)$.*

The intuition here is that most $F(q)$ will be zero, but once we see $F(q) > 0$ at some stage (when $f(q, s) > 0$), then we "know" its relationship with all those q' with $F(q') > 0$.

Theorem 5.2 (Kach and Turetsky [20]). *A set S has a computable increasing η-representation (with only finitely many blocks of any size $n > 1$) if and only if S is support increasing $\mathbf{0}'$-limitwise monotonic on $\mathbb{Q}$.*

Proof (Sketch). The forward direction is clear since given a $\mathbf{0}'$-oracle we know the blocks (monotonically) and how to order them. Of course, within the construction we might think that we have (distinct) blocks around q_1 and q_2. At a later stage, we may see these blocks merge, causing us to have an "extra" column of F which is positive. This is easily remedied by having both $h(q_1, s)$ and $h(q_2, s)$ reflect the merged size for all later s.

For the reverse direction, the construction proceeds as in Theorem 4.3. The only difference is that whenever we see $F(q) > 0$ for some new q (i.e., we see $f(q, s) > 0$), we create a new finite block within the linear order at the appropriate place. $\square$

It would be nice if altering the domain to $\mathbb{Q}$ would have application in characterizing the sets with computable strong η-representations. Frolov and Zubkov (see [13]) and Kach and Turestky (see [20]) have shown that there is a support increasing $\mathbf{0}'$-limitwise monotonic set on $\mathbb{Q}$ not having a computable strong η-representation. We finish this section by showing that being support strictly increasing $\mathbf{0}'$-limitwise monotonic on $\mathbb{Q}$ is not necessary to have a computable strong η-representation.

Theorem 5.3 (Turetsky). *There is a set S with a computable strong η-representation that is not support strictly increasing $\mathbf{0}'$-limitwise monotonic on $\mathbb{Q}$.*

Proof. Let $\{f_i(x, s)\}_{i \in \omega}$ be an enumeration of candidate total $\mathbf{0}'$-computable monotonic approximations on $\mathbb{Q}$. By the Limit Lemma, let $\{\hat{f}_i(x, s, t)\}_{i \in \omega}$ be an enumeration of computable approximations to f_i so that $f_i(x, s) = \lim_t \hat{f}_i(x, s, t)$. Note that since the f_i are total, the limit $\lim_t \hat{f}_i(x, s, t)$ will always converge.

We construct a computable presentation of a strong η-representation and let S be the set represented. We meet the following requirements:

$$\mathcal{R}_i : \text{The set } S \text{ is not the range of } F_i.$$

The strategy to assure $\mathcal{R}_i$ hinges on the fact that support strictly increasing limitwise monotonic functions cannot cope with two blocks in a strong η-representation merging. This fact is exploited to force a column to infinity.

Strategy for $\mathcal{R}_i$: Let $<_{\mathbb{Q}}$ be the natural ordering on $\mathbb{Q}$.

(1) Choose a large number n_0 and create blocks B_0 and B of sizes $n_0 - 1$ and n_0 in $\mathcal{L}$ at an appropriate location. Restrain other strategies from changing these blocks.
(2) Wait for a (least) pair $\langle x, u_0 \rangle$ to appear with $\hat{f}_i(x, u_0, t) = n_0$.
(3) Wait for a (least) pair $\langle x_0, s_0 \rangle$ to appear with $\hat{f}_i(x_0, s_0, t) = n_0 - 1$ and $x_0 <_{\mathbb{Q}} x$.
(4) Merge B_0 and B and any existing larger blocks into a single block of some size m_0 and release any restraint on this block. Restrain any blocks from forming of sizes between $n_0 - 1$ and m_0.

(5) Wait for an $s_0' > s_0$ with $\hat{f}_i(x_0, s_0', t) = m_0'$ for some $m_0' \geq m_0$. If more than one such s_0' exist, choose the least.
(6) Release the restraint created at Step 4.
(7) Wait for a $u_1 > u_0$ with $\hat{f}_i(x, u_1, t) = n_1$ for some $n_1 > m_0$ with n_1 the size of a block in $\mathcal{L}$.
(8) Create a block B_1 of size $n_1 - 1$ and restrain other strategies from changing this block or the block found in the previous step. Return to Step 3 with n_1 instead of n_0.

Note that our actions in Step 4 and Step 8 can be undone — we can resume densifying the interval between B_0 and B to separate the blocks, and we can densify the block B_1 to destroy it. Indeed, this capacity is essential, since there will be times we will need to roll back the construction to an earlier point. If, on some pair we chose, $\hat{f}_i$ changes its value, we return to the step at which we chose it, undoing all work done in the interim.

Thus, if at some stage t, $\hat{f}_i(x, u_0, t) \neq n_0$, we roll back the construction to Step 2. If at some stage t, $\hat{f}_i(x_j, s_j, t) \neq n_j - 1$, we roll back the construction to Step 3 in the jth loop. If at some stage t, $\hat{f}_i(x_j, s_j', t) \neq m_j'$, we roll back the construction to Step 5 in the jth loop, reestablishing the appropriate restraint. If at some stage t, $\hat{f}_i(x, u_j, t) \neq n_j$ (for $j > 0$), we roll back the construction to Step 7 in the jth loop.

Outcomes for $\mathcal{R}_i$: There are several possible outcomes for the strategy:

2: The strategy is infinitely often at Step 2, either because it waits at this step forever, or because it is infinitely often rolled back to this step. In either case, n_0 does not appear in the range of F_i but does appear as a block size in $\mathcal{L}$, and thus F_i does not enumerate S.

$\langle \mathbf{3}, \mathbf{j} \rangle$: The strategy is infinitely often at Step 3 in the jth loop, either because it waits at this step forever, or because it is infinitely often rolled back to this step. Further, none of outcomes **2**, $\langle \mathbf{3}, \mathbf{j'} \rangle$, $\langle \mathbf{5}, \mathbf{j'} \rangle$ or $\langle \mathbf{7}, \mathbf{j'} \rangle$ with $j' < j$ apply. In this case, $n_j - 1$ does not appear in the range of F_i but does appear as a block size in $\mathcal{L}$, and thus F_i does not enumerate S.

$\langle \mathbf{5}, \mathbf{j} \rangle$: The strategy is infinitely often at Step 5 in the jth loop, either because it waits at this step forever, or because it is infinitely often rolled back to this step. Further, none of outcomes **2**, $\langle \mathbf{3}, \mathbf{j'} \rangle$ with $j' \leq j$, or $\langle \mathbf{5}, \mathbf{j'} \rangle$ or $\langle \mathbf{7}, \mathbf{j'} \rangle$ with $j' < j$ apply. In this case, if $F_i(x_j)$ converges, then $F_i(x_j)$ is between $n_j - 1$ and m_j. However, S will have no element between $n_j - 1$ and m_j, and thus F_i does not enumerate S.

$\langle \mathbf{7,j} \rangle$: The strategy is infinitely often at Step 7 in the jth loop, either because it waits at this step forever, or because it is infinitely often rolled back to this step. Further, none of outcomes $\mathbf{2}$, $\langle \mathbf{3,j'} \rangle$ or $\langle \mathbf{5,j'} \rangle$ with $j' \leq j$, or $\langle \mathbf{7,j'} \rangle$ with $j' < j$ apply. Then if $F_i(x)$ converges, it does so to a value not contained in S. Thus F_i does not enumerate S.

∞: The strategy spends only finitely many stages at every step in every loop. Since $F_i(x) \geq n_j$ for all j, and $n_j < m_j < n_{j+1}$, $F_i(x)$ diverges.

The Tree: We order the outcomes of a strategy by:

$$\infty > \cdots > \langle \mathbf{7,1} \rangle > \langle \mathbf{5,1} \rangle > \langle \mathbf{3,1} \rangle > \langle \mathbf{7,0} \rangle > \langle \mathbf{5,0} \rangle > \langle \mathbf{3,0} \rangle > \mathbf{2}$$

As usual for infinite injury arguments, the *true outcome* of a strategy is the limit infimum of the outcomes.

We arrange the strategies on a tree in the usual fashion. When a strategy τ is rolled back, we also roll back the work done by any strategies ρ directly below τ.

If strategy ρ is below some finite outcome of strategy τ, the strategy ρ chooses a large n_0 and works with values larger than those used by τ. It is possible that ρ will be injured by a later merge step of τ. However, if we return to ρ, it will mean we have rolled back τ to before the merger, thus healing the injury to ρ.

If strategy ρ is below the infinite outcome of strategy τ, the strategy ρ waits for the restraint of τ to move to a sufficiently late interval that there is sufficient room for ρ to work with values beneath the restraint. It chooses its n_0 smaller than the restraint of τ, but larger than the current size of any blocks which existed when ρ was initialized. When ρ wishes to perform a merger, it waits until τ reaches a Step 6. It then performs the merger as described, including merging larger blocks that τ previously used. If at some later point τ is rolled back, the strategy ρ is rolled back with it.

If ρ is below the infinite outcome of τ, it is possible that τ will violate the restraint of ρ (if τ's n_j is ρ's m_k). In this case, ρ waits until τ performs a merger, and then reassigns m_k to the value of this new block (so ρ's m_k is τ's m_j). Barring roll back, τ will never again violate this restraint.

In this fashion, strategies respect the restraints imposed by strategies directly above them in the tree. Strategies pay no attention to restraints of any other strategies.

Verification: Define the True Path inductively using the limit infimum of the temporary outcomes.

Claim 5.3.1. If τ is along the True Path, and τ is active at stage t and has a restraint at stage t, then that restraint is not currently violated by some ρ directly below τ.

Proof. If ρ is below some finite outcome of τ, it creates blocks of size larger than the restraint of τ. If ρ is below the infinite outcome of τ, it respects the restraint of τ as discussed above. □

Claim 5.3.2. If τ is along the True Path, and τ is active at stage t and has a restraint at stage t, then that restraint is not currently violated by some ρ off the True Path.

Proof. Note that the restraint is not violated at the stage it is originally imposed.

Assume ρ is not directly below τ, as that case is handled above.

If the True Path follows a finite outcome at the first place it and ρ differ, and ρ is to the left of the True Path, then any activity by ρ between the stage at which the restraint is imposed and the current stage has been rolled back.

If the True Path follows a finite outcome at the first place it and ρ differ, and ρ is to the right of the True path, then ρ cannot act between the stage at which the restraint is imposed and the current stage (as in order for it to act, τ would have to be rolled back, removing the restraint).

If the True Path follows an infinite outcome at the first place it and ρ differ, then let σ be the meet of τ and ρ. Then ρ created blocks above the restraint of σ, while τ imposes its restraint beneath that of σ. □

Claim 5.3.3. If τ is along the True Path, and τ imposes a restraint, there will come a stage t when either τ will be rolled back to before it imposed this restraint, τ will release this restraint and this release will never be rolled back, or the restraint will never be violated after stage t.

Proof. Suppose that the restraint is neither rolled back nor released by τ. Then τ will wait until the σ above it stop violating the restraint. The strategy σ can only violate the restraint of τ if τ extends the infinite outcome of σ, and if σ has infinite final outcome, it can only be rolled back to any given step finitely many times. Thus, eventually, σ will never again violate the restraint of τ. Since no other strategies are capable of violating the restraint of τ, the restraint is never again violated. □

Claim 5.3.4. For any block created in $\mathcal{L}$, the limit infimum of its size is finite.

Proof. Let B be some block created by some strategy τ.

Suppose ρ is some other strategy. Let σ be ρ meet τ. In order for ρ to affect B, either ρ is σ or ρ is below the infinite outcome of σ, and either τ is σ or τ is below the finite outcome of σ. But by our construction of how strategies below an infinite outcome behave, ρ must have been initialized before B was created.

Thus there are only finitely many ρ that can affect B. Further, barring roll back, each strategy will only affect a given block finitely many times. Thus either one of these strategies is infinitely often rolled back, in which case B is constantly returned to a given finite size, or the size of B stabilizes. □

Claim 5.3.5. There are blocks of arbitrarily large size in $\mathcal{L}$.

Proof. Let τ be a strategy along the True Path being initialized at stage t such that this initialization will never be rolled back. During initialization, τ creates a large block. Since τ will never have its initialization rolled back, this block will never be destroyed. It may be grown into a larger block, but by the above, some large block will result. Thus $\mathcal{L}$ has arbitrarily large blocks. □

Claim 5.3.6. Each strategy along the True Path meets its requirement.

Proof. Immediate from construction. □

This completes the proof.

6. Limitwise Monotonic Spectra

As seemingly all of the effective algebra results with limitwise monotonicity relativize, there is a connection between the degree spectra of a structure

$$\mathrm{DegSpec}(\mathcal{S}) := \{\mathbf{a} : \mathcal{S} \text{ is } \mathbf{a}\text{-computable}\}$$

and the limitwise monotonic spectra of a set.

Definition 6.1. If $S \subseteq \omega$ is any nonempty set, define the *limitwise monotonic spectrum of* S (denoted $\mathrm{LMSpec}(S)$) to be the set

$$\mathrm{LMSpec}(S) := \{\mathbf{a} : S \text{ is } \mathbf{a}\text{-limitwise monotonic}\}.$$

In this language, we can reinterpret some of the material in the preceding sections. Since every Σ^0_2 degree has a limitwise monotonic set, we have the following.

Proposition 6.2 (Folklore). *If* $\mathbf{a}$ *is* Σ_2^0, *then there is an* $S \in \mathbf{a}$ *with* $\mathbf{0} \in$ *LMSpec*(S).

Theorem 6.3 (Khoussainov, Nies, and Shore [23]). *There exists a* Δ_2^0 *set* S, *indeed a d.c.e. set* S, *with* $\mathbf{0} \notin$ *LMSpec*(S).

Corollary 6.4 (Hirschfeldt, R. Miller, and Podzorov [15]). *There exists a low* Δ_2^0 *set* S *with* $\mathbf{0} \notin$ *LMSpec*(S).

We also demonstrate some new results.

Proposition 6.5. *There is a set* S *and a minimal pair of degrees* $\mathbf{a}$ *and* $\mathbf{b}$ *with* $\mathbf{a}, \mathbf{b} \in$ *LMSpec*(S) *and* $\mathbf{0} \notin$ *LMSpec*(S).

Proof. It suffices to fix a minimal pair of high degrees $\mathbf{a}$ and $\mathbf{b}$. Then $\mathbf{a}, \mathbf{b} \in$ LMSpec$(\emptyset''' \oplus \omega)$ by Theorem 2.1 (relativized) as $\emptyset''' \in \Sigma_3^0$ and $\Sigma_2^0(\mathbf{a}) = \Sigma_3^0 = \Sigma_2^0(\mathbf{b})$. On the other hand, it must be the case that $\mathbf{0} \notin$ LMSpec$(\emptyset''' \oplus \omega)$ as $\emptyset''' \oplus \omega \notin \Sigma_2^0$. □

Proposition 6.6 (Zubkov). *There are sets* S *and* T *with* $\mathbf{0} \in$ *LMSpec*(S), *LMSpec*(T) *and* $\mathbf{0} \notin$ *LMSpec*$(S \cap T)$.

Proof. By Theorem 2.2, fix a set S that is Σ_2^0 but not limitwise monotonic. Then $S \oplus \omega \oplus \emptyset$ and $S \oplus \emptyset \oplus \omega$ are limitwise motonic but their intersection is not. □

Proposition 6.7. *The containment LMSpec*$(T) \subseteq$ *LMSpec*(S) *does not follow from* $S \leq_T T$.

Proof. By Theorem 2.2, there is a Δ_2^0 set that is not a $\mathbf{0}$-limitwise monotonic set, yet $\emptyset'$ is a $\mathbf{0}$-limitwise monotonic set. □

Theorem 6.8. *If* $\mathbf{a}$ *and* $\mathbf{b}$ *satisfy* $\mathbf{a} < \mathbf{b}$, *then there is a set* S *with* $\mathbf{b} \in$ *LMSpec*(S) *and* $\mathbf{a} \notin$ *LMSpec*(S).

Proof. We start by noting that we may restrict attention to the case when $\mathbf{b} \in \Delta_2^0(\mathbf{a})$. For if $\mathbf{b} \notin \Delta_2^0(\mathbf{a})$, then (with $B \in \mathbf{b}$) either the set B or $\overline{B}$ suffices. The reason is both are clearly $\mathbf{b}$-limitwise monotonic. If both are $\mathbf{a}$-limitwise monotonic, then both are $\Sigma_2^0(\mathbf{a})$ by Theorem 2.1 (relativized). Being complements of each other, this implies both are $\Delta_2^0(\mathbf{a})$, contrary to the hypothesis.

Fix a total B-computable function g not dominated by any total A-computable function. We build F, a B-computable limitwise monotonic function, such that range(F) is not the range of any A-computable limitwise monotonic function Φ_e^A. We describe the construction relative to B.

We meet the following requirements:

$$\mathcal{R}_e : \text{range}(F) \neq \text{range}(\Phi_e^A)$$

The strategy to meet a single requirement in isolation is straightforward.

(1) Choose an integer z and define $f(e,s) := z$. Keep $F(y) \neq z$ for $y > e$.
(2) Wait for $\varphi_e^A(n,s) = z$ for some column n. While we wait, continue defining $f(e,s) := z$ for the current stage s.
(3) If $\varphi_e^A(n, g(z)) > z$, proceed to Step 7.
(4) Otherwise, define $f(e,s) := z+1$.
(5) Wait for $\varphi_e^A(n,s) > z$. While we wait, continue defining $f(e,s) := z+1$.
(6) Keep $F(y) \neq z+1$ for $y > e$. Release the z restraint, and return to Step 3 with $z+1$ in place of z.
(7) Wait for $\varphi_e^A(n,s) = f(w,s)$ for some $w > e$ and some $s > g(z)$. While we wait, continue defining $f(e,s) := z$.
(8) Return to Step 3 with w in place of e for the column of f and $f(w,s)$ in place of z.

Of course, if the strategy spends cofinitely many stages in Step 2, then $\mathcal{R}_i$ is satisfied as $z \in \text{range}(F)$ and $z \notin \text{range}(\Phi_e^A)$. If the strategy spends cofinitely many stages in Step 5, then $\Phi_e^A(n) \notin \text{range}(F)$. If the strategy spends cofinitely many stages in Step 7, then $\Phi_e^A(n) \notin \text{range}(F)$. If the strategy reaches Step 8 infinitely many times, then $\Phi_e^A(n)$ is infinite. If the strategy reaches Step 6 infinitely may times, then $\Phi_e^A(n)$ is infinite and the A-computable function $z \mapsto (\mu s)[\varphi_e^A(n,s) > z]$ dominates g, contrary to hypothesis.

As with Theorem 2.2, the strategies combine without any difficulty on a tree. Lower priority strategies guess whether the outcome of higher priority strategies is finitary (i.e., $\Phi_e(n)$ is finite) or infinitary (i.e., $\Phi_e(n)$ is infinite) □

7. Open Questions

We close with several questions (asked in numerous other places) that remain open.

Question 7.1. For which sets S is the strong η-representation of S computable?

Question 7.2. Which reduced abelian p-groups are computable? In particular, is there a reduced abelian p-group $\mathcal{G}$ of length ω^2 for which there is no $\mathbf{0}^{(2n)}$-computable approximation function $f(i,x,s)$ with $u_{\omega \cdot i + F(i,x)}(\mathcal{G}) > 0$?

Question 7.3. What more can be said about possible limitwise monotonic spectra?

References

[1] Klaus Ambos-Spies, S. Barry Cooper, and Steffen Lempp. Initial segments of recursive linear orders. *Order*, 14(2):101–105, 1997/98.

[2] C. J. Ash and J. Knight. *Computable structures and the hyperarithmetical hierarchy*, volume 144 of *Studies in Logic and the Foundations of Mathematics*. North-Holland Publishing Co., Amsterdam, 2000.

[3] William W. Boone. Certain simple, unsolvable problems of group theory. I. *Nederl. Akad. Wetensch. Proc. Ser. A.*, 57:231–237, 1954.

[4] William W. Boone. Certain simple, unsolvable problems of group theory. II. *Nederl. Akad. Wetensch. Proc. Ser. A.*, 57:492–497, 1954.

[5] William W. Boone. Certain simple, unsolvable problems of group theory. III. *Nederl. Akad. Wetensch. Proc. Ser. A.*, 58:252–256, 1955.

[6] William W. Boone. Certain simple, unsolvable problems of group theory. IV. *Nederl. Akad. Wetensch. Proc. Ser. A.*, 17:571–577, 1955.

[7] William W. Boone. Certain simple, unsolvable problems of group theory. V, VI. *Nederl. Akad. Wetensch. Proc. Ser. A.*, 19:22–27, 227–232, 1957.

[8] Wesley Calvert, Douglas Cenzer, Valentina Harizanov, and Andrei Morozov. Effective categoricity of equivalence structures. *Ann. Pure Appl. Logic*, 141(1-2):61–78, 2006.

[9] Richard J. Coles, Rod Downey, and Bakhadyr Khoussainov. On initial segments of computable linear orders. *Order*, 14(2):107–124, 1997/98.

[10] R. G. Downey. Computability theory and linear orderings. In *Handbook of recursive mathematics, Vol. 2*, volume 139 of *Studies in Logic and the Foundations of Mathematics*, pages 823–976. North-Holland, Amsterdam, 1998.

[11] Rod Downey, Bakhadyr Khoussainov, Jospeh S. Miller, and Liang Yu. Degree spectra of unary relations on $\langle \omega, \leq \rangle$. In *Logic, Methodology and Philosophy of Science*, Proceedings of the Thirteenth International Congress, pages 36–65. King's College Publications.

[12] L. Feiner. The strong homogeneity conjecture. *J. Symbolic Logic*, 35:375–377, 1970.

[13] Andrey N. Frolov and Maxim V. Zubkov. Increasing η-representable degrees. *Math. Log. Q.*, 55(6):633–636, 2009.

[14] Kenneth Harris. η-representation of sets and degrees. *J. Symbolic Logic*, 73(4):1097–1121, 2008.

[15] Denis Hirschfeldt, Russell Miller, and Sergei Podzorov. Order-computable sets. *Notre Dame J. Formal Logic*, 48(3):317–347 (electronic), 2007.

[16] Denis R. Hirschfeldt. Prime models of theories of computable linear orderings. *Proc. Amer. Math. Soc.*, 129(10):3079–3083 (electronic), 2001.

[17] Denis R. Hirschfeldt, Bakhadyr Khoussainov, and Pavel Semukhin. An uncountably categorical theory whose only computably presentable model is saturated. *Notre Dame J. Formal Logic*, 47(1):63–71 (electronic), 2006.

[18] Asher M. Kach. Computable shuffle sums of ordinals. *Archive for Mathematical Logic*, 47(3):211–219, 2008.

[19] Asher M. Kach and Joseph S. Miller. Embeddings of computable linear orders. In preparation.

[20] Asher M. Kach and Daniel Turetsky. Limitwise monotonic functions, sets, and degrees on computable domains. *J. Symbolic Logic*, 75(1):131–154, 2010.

[21] N. G. Khisamiev. The arithmetic hierarchy of abelian groups. *Sibirsk. Mat. Zh.*, 29(6):144–159, 1988.

[22] N. G. Khisamiev. Constructive abelian groups. In *Handbook of recursive mathematics, Vol. 2*, volume 139 of *Stud. Logic Found. Math.*, pages 1177–1231. North-Holland, Amsterdam, 1998.

[23] Bakhadyr Khoussainov, Andre Nies, and Richard A. Shore. Computable models of theories with few models. *Notre Dame J. Formal Logic*, 38(2):165–178, 1997.

[24] Carolyn Knoll. *Degree Spectra of Unary Relations on* $(\omega, \leq)$ *and* $(\zeta, \leq)$. M.S. in Mathematics, University of Waterloo, http://hdl.handle.net/10012/4544, 2009.

[25] Manuel Lerman. On recursive linear orderings. In *Logic Year 1979–80 (Proc. Seminars and Conf. Math. Logic, Univ. Connecticut, Storrs, Conn., 1979/80)*, volume 859 of *Lecture Notes in Math.*, pages 132–142. Springer, Berlin, 1981.

[26] P. S. Novikov. *Ob algoritmičeskoĭ nerazrešimosti problemy toždestva slov v teorii grupp.* Trudy Mat. Inst. im. Steklov. no. 44. Izdat. Akad. Nauk SSSR, Moscow, 1955.

[27] Emil L. Post. Recursively enumerable sets of positive integers and their decision problems. *Bull. Amer. Math. Soc.*, 50:284–316, 1944.

[28] Emil L. Post. Recursive unsolvability of a problem of Thue. *J. Symbolic Logic*, 12:1–11, 1947.

[29] Matthew James S. Raw. *Complexity of Automorphisms of Recursive Linear Orders.* PhD in Mathematics, University of Wisconsin-Madison, 1995.

[30] Joseph G. Rosenstein. *Linear orderings*, volume 98 of *Pure and Applied Mathematics.* Academic Press Inc. [Harcourt Brace Jovanovich Publishers], New York, 1982.

[31] Maxim Zubkov. On η-representable sets. In *Computation and logic in the real world*, pages 364–366. 2007.

A DICHOTOMY FOR THE MACKEY BOREL STRUCTURE

Ilijas Farah

Department of Mathematics and Statistics
York University, 4700 Keele Street, North York, Ontario, Canada, M3J 1P3
Matematički Institut, Kneza Mihaila 34, 11 000 Beograd, Serbia
http://www.math.yorku.ca/~ifarah
ifarah@mathstat.yorku.ca

We prove that the equivalence of pure states of a separable C*-algebra is either smooth or it continuously reduces $[0,1]^{\mathbb{N}}/\ell_2$ and it therefore cannot be classified by countable structures. The latter was independently proved by Kerr–Li–Pichot by using different methods. We also give some remarks on a 1967 problem of Dixmier.

1991 *Mathematics Subject Classification*: 03E15, 46L30, 22D25.
Keywords and phrases: Borel equivalence relations, Mackey Borel structure.

If E and F are Borel equivalence relations on Polish spaces X and Y, respectively, then we say that E is *Borel reducible* to F (in symbols, $E \leq_B F$) if there is a Borel-measurable map $f\colon X \to Y$ such that for all x and y in X we have xEy if and only if $f(x)Ff(y)$. A Borel equivalence relation E is *smooth* if it is Borel-reducible to the equality relation on some Polish space. Recall that E_0 is the equivalence relation on $2^{\mathbb{N}}$ defined by xE_0y if and only if $x(n) = y(n)$ for all but finitely many n. The *Glimm–Effros* dichotomy ([8]) states that a Borel equivalence relation E is either smooth or $E_0 \leq_B E$.

One of the themes of the abstract classification theory is measuring relative complexity of classification problems from mathematics (see e.g., [12]). One can formalize the notion of 'effectively classifiable by countable structures' in terms of the relation $\leq_B$ and a natural Polish space of structures based on $\mathbb{N}$ in a natural way. In [10] Hjorth introduced the notion of turbu-

The work reported in this note was done in July 2008 while I was visiting IHES and it was presented at a mini-conference in set theory at the Institut Henri Poincaré in July 2008.

lence for orbit equivalence relations and proved that an orbit equivalence relation given by a turbulent action cannot be effectively classified by countable structures.

The idea that there should be a small set $\mathcal{B}$ of Borel equivalence relations not classifiable by countable structures such that for every Borel equivalence relation E not classifiable by countable structures there is $F \in \mathcal{B}$ such that $F \leq_B E$ was put forward in [11] and, in a revised form, in [4]. In this note we prove a dichotomy for a class of Borel equivalence relations corresponding to the spectra of C*-algebras by showing that one of the standard turbulent orbit equivalence relations, $[0,1]^{\mathbb{N}}/\ell_2$, is Borel-reducible to every non-smooth spectrum.

States

All undefined notions from the theory of C*-algebras and more details can be found in [2] or in [5]. Consider a separable C*-algebra A. Recall that a functional ϕ on A is *positive* if it sends every positive operator in A to a positive real number. A positive functional is a *state* if it is of norm ≤ 1. The states form a compact convex set, and the extreme points of this set are the *pure states*. The space of pure states on A, denoted by $\mathbb{P}(A)$, equipped with the weak*-topology, is a Polish space ([14, 4.3.2]).

A C*-algebra A is *unital* if it has the multiplicative identity. Otherwise, we define the *unitization* of A, $\tilde{A}$, the canonical unital C*-algebra that has A as a maximal ideal and such that the quotient $\tilde{A}/A$ is isomorphic to $\mathbb{C}$ (see [5, Lemma 2.3]). If u is a unitary in A (or $\tilde{A}$) then

$$(\operatorname{Ad} u)a = uau^*$$

defines an inner automorphism of A.

Two pure states ϕ and ψ are equivalent, $\phi \sim_A \psi$, if there exists a unitary u in A (or $\tilde{A}$) such that $\phi = \psi \circ \operatorname{Ad} u$.

Theorem 1. *Assume A is a separable C*-algebra. Then $\sim_A$ is either smooth or there is a continuous map*

$$\Phi\colon [0,1]^{\mathbb{N}} \to \hat{A}$$

such that $\alpha - \beta \in \ell_2$ if and only if $\Phi(\alpha) \sim_A \Phi(\beta)$.

Corollary 2. *Assume A is a separable C*-algebra. Then either $\sim_A$ is smooth or it cannot be classified by countable structures.*

Proof. By [10] it suffices to show that a turbulent orbit equivalence relation is Borel-reducible to $\sim_A$ if $\sim_A$ is not smooth. The equivalence relation

$[0,1]^{\mathbb{N}}/\ell_2$ is well-known to be turbulent (e.g., [11]) and the conclusion follows by Theorem 1. □

This result was independently proved in [13, Theorem 2.8] by directly showing the turbulence. As pointed out in [13, §3], it implies an analogous result of Hjorth ([9]) on irreducible representations of discrete groups, as well as its strengthening to locally compact groups.

1. Proof of Theorem 1

Recall that the CAR (Canonical Anticommutation Relations) algebra (also know as the Fermion algebra, or M_{2^∞}) is defined as the infinite tensor product

$$M_{2^\infty} = \bigotimes_{n\in\mathbb{N}} M_2(\mathbb{C})$$

where $M_2(\mathbb{C})$ is the algebra of 2×2 matrices. Alternatively, one may think of M_{2^∞} as the direct limit of $2^n \times 2^n$ matrix algebras $M_{2^n}(\mathbb{C})$ for $n\in\mathbb{N}$.

The following analogue of the Glimm–Effros dichotomy is an immediate consequence of [6] (Notably, the key combinatorial device in the proof of [8] comes from Glimm).

Proposition 3. *If A is a separable C^*-algebra then exactly one of the following applies.*

(1) *$\sim_A$ is smooth.*
(2) *$\sim_{M_{2^\infty}} \leq_B \sim_A$.* □

We shall prove that $\sim_{M_{2^\infty}}$ is turbulent in the sense of Hjorth.

Lemma 4. *If ξ and η are unit vectors in H then*

$$\inf\{\|I-u\| : u \text{ unitary in } \mathcal{B}(H) \text{ and } (u\xi|\eta)=1\} = \sqrt{2(1-|(\xi|\eta)|)}. \quad (*)$$

Proof. Let $t=(\xi|\eta)$. Let $\xi' = \frac{1}{\|\operatorname{proj}_{\mathbb{C}\eta}\xi\|}\operatorname{proj}_{\mathbb{C}\eta}\xi$. Then the square of the left-hand side of (*) is greater than or equal to"

$$\|\xi-\xi'\|^2 = \|\xi\|^2 + \|\xi'\|^2 - \frac{2}{(\xi|\eta)}(\xi|(\xi|\eta)\eta) = 2-2|t|.$$

For $\leq$ let ζ be the unit vector orthogonal to ξ such that

$$\eta = t\xi + \sqrt{1-t^2}\zeta$$

and let u be the unitary given by $\begin{pmatrix} t & -\sqrt{1-t^2} \\ \sqrt{1-t^2} & t \end{pmatrix}$ on the span of ξ and ζ and identity on its orthogonal complement. Then $u\xi = \eta$ and a straightforward computation gives $\|I - u\|^2 = 2 - 2t$ as required. $\square$

If ξ is a unit vector in a Hilbert space then by ω_ξ we denote the vector state $a \mapsto (a\xi|\xi)$. If ξ_i is a unit vector in H_i for $1 \leq i \leq m$ then $\xi = \bigotimes_{i=1}^m \xi_i$ is a unit vector in $H = \bigotimes_{i=1}^m \xi_i$ and ω_ξ is a vector state on $\mathcal{B}(H)$.

Lemma 5. *If H_i is a Hilbert space and ξ_i, η_i are unit vectors in H_i for $1 \leq i \leq m$ then*

$$\inf\{\|I-u\| : u \text{ unitary and } \omega_{\bigotimes_{i=1}^m \xi_i} = \omega_{\bigotimes_{i=1}^m \eta_i} \circ \operatorname{Ad} u\} = 2\sqrt{2(1 - \prod_{i=1}^m |(\xi_i|\eta_i)|)}.$$

Proof. The case when $m = 1$ follows from Lemma 4 and the fact that $\omega_\xi = \omega_{\alpha\xi}$ when $|\alpha| = 1$. Since $(\bigotimes_{i=1}^m \xi_i | \bigotimes_{i=1}^m \eta_i) = \prod_{i=1}^m (\xi_i|\eta_i)$, the general case is an immediate consequence of Lemma 4. $\square$

Theorem 6. *There is a continuous map $\Phi\colon (-\frac{\pi}{2}, \frac{\pi}{2})^{\mathbb{N}} \to \mathbb{P}(M_{2^\infty})$ such that for all $\vec{\alpha}$ and $\vec{\beta}$ in the domain we have*

$$\sum_n (\alpha_n - \beta_n)^2 < \infty \Leftrightarrow \Phi(\vec{\alpha}) \sim_{M_{2^\infty}} \Phi(\vec{\beta}).$$

Proof. Consider the standard representation of $M_2(\mathbb{C})$ on $\mathbb{C}^2$. Then the pure states of $M_2(\mathbb{C})$ are of the form $\omega_{(\cos\alpha, \sin\alpha)}$ for $\alpha \in (-\frac{\pi}{2}, \frac{\pi}{2})$.

Let $\Phi(\vec{\alpha}) = \bigotimes_{n=1}^\infty \omega_{(\cos\alpha_n, \sin\alpha_n)}$. This map is continuous: If $a \in M_{2^\infty}$ and $\varepsilon > 0$, fix m and $a' \in M_{2^m}$ such that $\|a - a'\| < \varepsilon/2$. Then $\Phi(\vec{\alpha})(a')$ depends only on α_j for $j \leq m$, and in a continuous fashion.

Recall that for $0 < t_j < 1$ we have $\prod_{j=1}^\infty t_j > 0$ if and only if $\sum_{j=1}^\infty (1 - t_j) < \infty$. Therefore

$$\sum_{n=1}^\infty (\alpha_n - \beta_n)^2 < \infty \Leftrightarrow \sum_{n=1}^\infty \sin^2\left(\frac{\alpha_n - \beta_n}{2}\right) < \infty \Leftrightarrow \prod_{n=1}^\infty \cos(\alpha_n - \beta_n) > 1.$$

Assume $\prod_{n=1}^\infty \cos(\alpha_n - \beta_n) > 0$. In the n-th copy of M_2 in $M_{2^\infty} = \bigotimes_{n=1}^\infty M_2$ pick a unitary u_n such that

$$\|1 - u_n\| < \sqrt{2(1 - |\cos(\alpha_n - \beta_n)|)}$$

and $u_n(\cos\alpha_n, \sin\alpha_n) = (\cos\beta_n, \sin\beta_n)$. Note that

$$((\cos\alpha_n, \sin\alpha_n)|(\cos\beta_n, \sin\beta_n)) = \cos(\alpha_n - \beta_n).$$

Let $v_n = \bigotimes_{j=1}^n u_j$. Then v_n for $n \in \mathbb{N}$ form a Cauchy sequence, because $v_m - v_{m+n} = v_m(1 - \bigotimes_{j=m+1}^n u_j)$ and therefore

$$\|v_m - v_n\| < \sqrt{2(1 - \prod_{j=m}^\infty \cos(\alpha_j - \beta_j))}.$$

Let $v \in M_{2^\infty}$ be the limit of this Cauchy sequence. Since for each m and $a \in M_{2^m}$ we have $\Phi(\vec{\alpha})(a) = \Phi(\vec{\beta})(v_n a v_n^*)$ for any $n \geq m$, we have $\Phi(\vec{\alpha}) = \Phi(\vec{\beta}) \circ \operatorname{Ad} v$.

Now assume $\Phi(\vec{\alpha}) \sim_{M_{2^\infty}} \Phi(\vec{\beta})$ and, for the sake of obtaining a contradiction, that $\prod_{n=1}^\infty \cos(\alpha_n - \beta_n) = 0$. There is m and a unitary $u \in M_{2^m}$ such that

$$\|\Phi(\vec{\alpha}) - \Phi(\vec{\beta}) \circ \operatorname{Ad} u\| < \frac{1}{2}.$$

(by e.g., [6]). However, we can find $n > m$ large enough so that with $\xi_n = \bigotimes_{j=m}^n (\cos\alpha_j, \sin\alpha_j)$ and $\eta_n = \bigotimes_{j=m}^n (\cos\beta_j, \sin\beta_j)$ the quantity

$$(\xi_n|\eta_n) = \prod_{j=m}^n \cos(\alpha_j - \beta_j)$$

is as close to zero as desired. Then $\|\omega_{\xi_n} - \omega_{\eta_n}\|$ is as close to 2 as desired, since $a_n = \operatorname{proj}_{\mathbb{C}\xi_n} - \operatorname{proj}_{\mathbb{C}\eta_n}$ has norm close to 1 and $\omega_{\xi_n}(a_n)$ is close to 1 while $\omega_{\eta_n}(a_n)$ is close to -1. □

Proof of Theorem 1. Assume $\sim_A$ is not smooth. The conclusion follows by Glimm's Proposition 3 and Theorem 6. □

2. Concluding Remarks

We note that the class of equivalence relations corresponding to spectra of C*-algebras is restrictive in another sense. The following proposition was probably well-known (cf. [9, Corollary 1.3]).

Proposition 7. *If A is a separable C*-algebra then the relation $\phi \sim_A \psi$ on $\mathbb{P}(A)$ is F_σ.*

Proof. By replacing A with its unitization if necessary we may assume A is unital. Fix a countable dense set $\mathcal{U}$ in the unitary group of A and a countable dense set $\mathcal{D}$ in $A_{\leq 1}$. We claim that

$$\phi \sim_A \psi \Leftrightarrow (\exists u \in \mathcal{U})(\forall a \in \mathcal{D})|\phi(a) - \psi(uau^*)| < 1.$$

Assume $\phi \sim_A \psi$ and fix v such that $\phi = \psi \circ \operatorname{Ad} v$. If $u \in \mathcal{U}$ is such that $\|v - u\| < 1/2$ then

$$|\psi(uau^* - vav^*)| = |\psi((u - v)au^* - va(u^* - v^*))| < 1$$

for all $a \in A_{\leq 1}$.

Now assume $u \in \mathcal{U}$ is such that $|\phi(a) - \psi(uau^*)| < 1$ for all $a \in \mathcal{D}$. Then $\|\phi - \psi \circ \operatorname{Ad} u\| < 2$ and by [7] we have $\phi \sim_A \psi$. □

For a Hilbert space H by $\mathcal{B}(H)$ we denote the algebra of its bounded linear operators. Let $\pi_1 \colon A \to \mathcal{B}(H_1)$ and $\pi_2 \colon A \to \mathcal{B}(H_2)$ be representations of A. We say π_1 and π_2 are *(unitarily) equivalent* and write $\pi_1 \sim \pi_2$ if there is a Hilbert space isomorphism $u \colon H_1 \to H_2$ such that the diagram

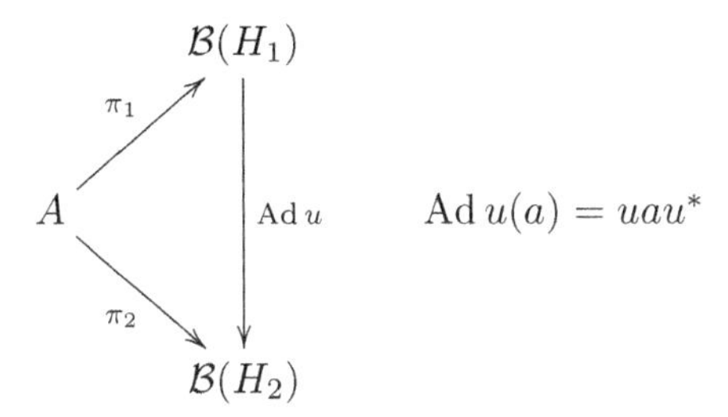

commutes.

A representation of A on some Hilbert space H is *irreducible* if there are no nontrivial closed subspaces of H invariant under the image of A. The spectrum of A, denoted by $\hat{A}$, is the space of all equivalence classes of irreducible representations of A. The GNS construction associates a representation π_ϕ of A to each state ϕ of A (see e.g., [5, Theorem 3.9]). Moreover, ϕ is pure if and only if π_ϕ is irreducible ([5, Theorem 3.12]) and for pure states ϕ_1 and ϕ_2 we have that ϕ_1 and ϕ_2 are equivalent if and only if π_{ϕ_1} and π_{ϕ_2} are equivalent ([5, Proposition 3.20]).

Fix a separable C*-algebra A. Let $\operatorname{Irr}(A, H_n)$ denote the space of irreducible representations of A on a Hilbert space H_n of dimension n for $n \in \mathbb{N} \cup \{\aleph_0\}$. Each $\operatorname{Irr}(A, H_n)$ is a Polish space with respect to the weakest topology making all functions $\operatorname{Irr}(A, H_n) \ni \pi \mapsto (\pi(a)\xi|\eta) \in \mathbb{C}$, for $a \in A$ and $\xi, \eta \in H_n$, continuous. In other words, a net π_λ converges to π if and only if $\pi_\lambda(a)$ converges to $\pi(a)$ for all $a \in A$. Since A is separable, each irreducible representation of A has range in a separable Hilbert space, and therefore $\hat{A}$ can be considered as a quotient space of the direct sum of $\operatorname{Irr}(A, H_n)$ for $n \in \mathbb{N} \cup \{\aleph_0\}$. Therefore $\hat{A}$ carries a Borel structure (known as the *Mackey Borel structure*) inherited from a Polish space. For *type I*

C*-algebras (also called *GCR* or *postliminal*) this space is a standard Borel space. (All of these notions are explained in [1, §4].)

Since pure states correspond to irreducible representations, we can identify the Mackey Borel structure of A with a σ-algebra of sets in $\hat{A}$. It is easy to check that this σ-algebra consists exactly of those sets whose preimages in $\mathbb{P}(A)$ are Borel subsets in $\mathbb{P}(A)$.

Glimm proved ([6], [14, §6.8]) that the Mackey Borel structure of a C*-algebra A is *smooth* (i.e., isomorphic to a standard Borel space) if and only if A is a type I C*-algebra. Proposition 3 is a consequence of this result.

Problem 8 (Dixmier, 1967). *Is the Mackey Borel structure on the spectrum of a simple separable C*-algebra always the same when it is not standard?*

G. Elliott generalized Glimm's result and proved that the Mackey Borel structures of simple AF algebras are isomorphic ([3]). (A C*-algebra is an AF (approximately finite) algebra if it is a direct limit of finite-dimensional algebras.) One reformulation of Elliott's result is that for any two simple separable AF algebras A and B there is a Borel isomorphism $F\colon \mathbb{P}(A) \to \mathbb{P}(B)$ such that $\phi \sim_A \psi$ if and only if $F(\phi) \sim_B F(\psi)$ (see [3, §6]). Also, [3, Theorem 2] implies that if A is a simple separable AF algebra and B is a non-Type I simple separable algebra we have $\sim_A \leq_B \sim_B$.

With this definition the quotient structure $\mathrm{Borel}(\mathbb{P}(A))/\sim_A$ is isomorphic to the Mackey Borel structure of A. Note that $\sim_A$ is smooth exactly when the Mackey Borel structure of A is smooth.

Note that Mackey Borel structures of A and B of separable C*-algebras are isomorphic if and only if there is a Borel isomorphism $f\colon \hat{X} \to \hat{X}$ such that $\pi_1 \sim_A \pi_2$ if and only if $f(\pi_1) \sim_B f(\pi_2)$. Hence Problem 8 is rather close in spirit to the theory of Borel equivalence relations.

N. Christopher Phillips suggested more general problems about the Mackey Borel structure of simple separable C*-algebras, motivated by his discussions with Masamichi Takesaki. There are two (related) kinds of questions: Can one do anything sensible, and, from the point of view of logic, how bad is the problem?

Problem 9. Does the complexity of the Mackey Borel structure of a simple separable C*-algebra increase as one goes from nuclear C*-algebras to exact ones to ones that are not even exact?

For definitions of nuclear and exact C*-algebras see e.g., [2].

Problem 10. Assume A and B are C*-algebras and $\sim_A$ is Borel-reducible to $\sim_B$. What does this fact imply about the relation between A and B?

References

[1] W. Arveson, *An invitation to C^*-algebras*, Springer-Verlag, New York, 1976, Graduate Texts in Mathematics, No. 39.

[2] B. Blackadar, *Operator algebras*, Encyclopaedia of Mathematical Sciences, vol. 122, Springer-Verlag, Berlin, 2006, Theory of C^*-algebras and von Neumann algebras, Operator Algebras and Non-commutative Geometry, III.

[3] G.A. Elliott, *The Mackey Borel structure on the spectrum of an approximately finite-dimensional separable C^*-algebra*, Trans. Amer. Math. Soc. **233** (1977), 59–68.

[4] I. Farah, *Basis problem for turbulent actions II: c_0-equalities*, Proceedings of the London Mathematical Society **82** (2001), 1–30.

[5] I. Farah and E. Wofsey, *Set theory and operator algebras*, Proceedings of the Appalachian set theory workshop. http://www.math.cmu.edu/~eschimme/Appalachian/Index.html (E. Schimmerling et al., ed.), to appear.

[6] J. G. Glimm, *On a certain class of operator algebras*, Trans. Amer. Math. Soc. **95** (1960), 318–340.

[7] J. G. Glimm and R. V. Kadison, *Unitary operators in C^*-algebras*, Pacific J. Math. **10** (1960), 547–556.

[8] L.A. Harrington, A.S. Kechris, and A. Louveau, *A Glimm–Effros dichotomy for Borel equivalence relations*, Journal of the American Mathematical Society **4** (1990), 903–927.

[9] G. Hjorth, *Non-smooth infinite dimensional group representations*, preprint, available at http://www.math.ucla.edu/~greg/, 1997.

[10] ———, *Classification and orbit equivalence relations*, Mathematical Surveys and Monographs, vol. 75, American Mathematical Society, 2000.

[11] G. Hjorth and A.S. Kechris, *New dichotomies for Borel equivalence relations*, The Bulletin of Symbolic Logic **3** (1997), 329–346.

[12] G Hjorth and A.S. Kechris, *Recent developments in the theory of borel reducibility*, Fundam. Math. **170** (2001), 21–52.

[13] D. Kerr, H. Li, and M. Pichot, *Turbulence, representations, and trace-preserving actions*, Proc. London Math. Soc. (to appear).

[14] Gert K. Pedersen, *C^*-algebras and their automorphism groups*, London Mathematical Society Monographs, vol. 14, Academic Press Inc. [Harcourt Brace Jovanovich Publishers], London, 1979.

ON AUTOMATIC FAMILIES

Sanjay Jain*, Yuh Shin Ong† and Shi Pu‡

School of Computing, National University of Singapore
Block COM1, 13 Computing Drive, Singapore 117417
**sanjay@comp.nus.edu.sg, †yuhshin@gmail.com and ‡pushi@nus.edu.sg*

Frank Stephan

Department of Mathematics, National University of Singapore
Block S17, 10 Lower Kent Ridge Road, Singapore 119076
fstephan@comp.nus.edu.sg

This paper summarises previous work on automatic families. It then investigates a natural size measure for members of an automatic family: the size of a member language in the family is defined as the length of its smallest index. This measure satisfies various properties similar to those of Kolmogorov complexity; in particular the size of a language depends only up to a constant on the underlying automatic family. This family of size measures is extended to a measure on all regular sets. This extension is given by the maximum number of states visited in some run of the minimal deterministic finite automaton recognising the language. Furthermore, a characterisation is given regarding when a class of languages is a subclass of an automatic family.

Sanjay Jain and Frank Stephan are supported in part by NUS grant numbers R252-000-308-112 and R146-000-114-112.

1. Introduction

Automatic families are uniformly regular families of sets which are mainly used in inductive inference (learning theory) as a way to represent hypothesis spaces with decidable first-order theory. For such spaces, one can decide whether they are explanatorily learnable in the limit and have effective procedures to generate such learners. Furthermore, they have also been used to study various other learnability notions. Automatic families are closely related to automatic graphs and automatic structures. All these notions are based on the notion of a finite automaton. The present work formalises the notion of a size of a language with respect to an automatic class. It also relates this size to various measures of the size of a language derived from its minimal deterministic finite automaton. Furthermore, an overview of related properties and applications of automatic families is given.

The basic notion of the field is that of a finite automaton. Informally, a finite automaton could be considered as an algorithm which reads a word from the left to the right and has a constant-sized memory; when it reaches the right end of the word, the algorithm says either "accept" or "reject". This informal description permits to write algorithms representing finite automata quite well; on the other hand, when one wants to prove properties of finite automata, it is better to write them as a finite set of states together with a state-transition table, a starting state and a set of final states. The automaton then reads the symbols of the word to be checked from the left to the right and each time changes its state based on a transition-table which is a set of triples of the form (old state, current symbol, new state). Ideally, for every old state and current symbol, there is a unique triple in the table; an automaton with this property is called a "deterministic finite automaton" (dfa) and every finite automaton can be converted into an equivalent automata of this form, which accepts the same language as the original automaton. Now, for studying automatic structures, the key generalisation is to permit automata to track various inputs simultaneously, provided that the reading of these inputs is synchronised. Informally, the algorithm consists of one loop. In each iteration of the loop the algorithm first reads, from each of those inputs which are not already exhausted, one symbol and then updates its internal memory accordingly; the memory is restricted to a constant size. Here, the algorithm knows when an input is exhausted and when not, so this piece of information can be taken into account. When every input is exhausted, the algorithm says "accept" or "reject". The fact that all inputs are read at the same speed is important, otherwise the model would become too powerful and certain undecidability problems of the theory of such models would arise. Mathematically, one can also map this back to the case of a single input which is formed as the convolution of the inputs. The convolution $conv(x, y)$ of two inputs x and y would consist of symbols which are pairs of the corresponding symbols from x and y; in the case that one of these input words is shorter than the other, it is brought up to the same length by appending the special symbol $\diamond$ sufficiently often; the symbol $\diamond$ is not contained in any of the input alphabets used for x and y. One can define $conv$ on more than two arguments similarly. The next algorithm is an example which decides the lexicographic order of two strings x and y; the algorithm uses an underlying ordering $<$ on Σ:

(1) If the input x is exhausted then accept and terminate.
(2) If the input y is exhausted then reject and terminate.
(3) Read the current symbol a of the input x and the current symbol b of the input y.
(4) If $a < b$ then accept and terminate.
(5) If $b < a$ then reject and terminate.
(6) Go to (1).

Note that this algorithm also accepts if both words are equal; so it accepts if $x \leq_{lex} y$ and rejects if $y <_{lex} x$. Furthermore, the algorithm produces an early decision, when possible. So when comparing 25880 with 2593 the algorithm would already make the decision after having read 258 from 25880 and 259 from 2593. Such an early decision is permitted for easier formulations of the algorithms, although formally the algorithm has to scan the full input.

Besides automatic relations like the lexicographic ordering, one can also define automatic functions where the automaton recognises a function f as follows: it reads convolution of x and y and accepts iff x is in the domain of f and $y = f(x)$. An automatic family is a structure with the following properties: One has a domain D and an index domain I which are both regular sets plus an automatic relation on $I \times D$ defining a family $\{L_i : i \in I\}$ such that $x \in L_i$ iff the underlying relation contains (i, x). Such a family defines a class of subsets of D and two families have the same range iff they contain the same languages. In many cases it is also handy to have that the indexing is a one-one indexing, that is, $L_i \neq L_j$ for different $i, j \in I$. However, this property is not mandatory within the present work. An important result from the early days of automatic structures [8, 9, 17] is that whenever a relation or function is first-order definable using other automatic relations or functions then it is itself automatic. Furthermore, the first-order theory of automatic structures is decidable. These two results give this field some importance in model checking and similar applications and the two results are also frequently used when constructing learners in inductive inference. The next examples of automatic families illustrate the concept.

Example 1.1: Let $\Sigma = \{0, 1, 2\}$ and $D = \Sigma^*$.

The family of all $L_i = \Sigma^{|i|}$, where $i \in \{0\}^*$, is an automatic family. The automaton recognising the family accepts $conv(i, x)$ iff i and x have the same length.

The family of all $L_i = \{x \in D : x \text{ extends } i\}$ with $i \in \Sigma^*$ is an automatic family; an example member of it is $L_{001} = \{001, 0010, 0011, 0012, 00100, 00101, \ldots, 00122, 001000, 001001, \ldots\}$. The automaton recognising the family accepts $conv(i, x)$ iff every symbol of i appears at the same position in x.

The family $L_{conv(i,j)} = \{x : i <_{lex} x <_{lex} j\}$ of all $i, j \in \Sigma^*$ with $i <_{lex} j$ is also automatic; note that in some special cases like $j = i0$ the language $L_{conv(i,j)}$ is empty. The automaton recognising the family is mainly checking whether x is between the two boundaries i and j which are coded up in the index $conv(i, j)$.

The family of all L_{i3a3b} with $i3a3b \in I = \Sigma^* \cdot \{3\} \cdot \Sigma \cdot \{3\} \cdot \Sigma$ and $L_{i3a3b} = \{ixaybz : x, y, z \in \Sigma^*\}$ is automatic. For example, $L_{00123231}$ is given by the regular expression $0012 \cdot (0 + 1 + 2)^* \cdot 2 \cdot (0 + 1 + 2)^* \cdot 1 \cdot (0 + 1 + 2)^*$. The automaton recognising the automatic family has to memorise the values of a and b when reaching the corresponding position in the input $conv(i3a3b, u)$ as a and b can occur in u much later than in $i3a3b$. This is possible as a finite automaton can memorise a constant amount of information.

Example 1.2: Using a certain regular domain D, one can also code the natural numbers with addition, the relation $<$ and a predicate Fib recognising the Fibonacci numbers [28]. If now $\phi(x, i, j, k)$ is a first-order formula with four inputs defined using $+$, $<$, Fib and natural numbers then the family of all $L_{conv(i,j,k)} = \{x \in D : \phi(x, i, j, k) \text{ is true}\}$ is an automatic family where the index-domain is the set of all convolutions of three elements of D. An example for such a formula ϕ is $\phi(x, i, j, k) \Leftrightarrow \exists y \exists z [i < x + y < j \wedge x + x = z + z + z + 2 \wedge Fib(x + y + y + k)]$.

Note that the above example is more in the traditional style of automatic structures where all aspects of coding can be freely chosen in order to meet the specification. Automatic families usually are a bit more fixed; here one wants to find an automatic indexing of a given class of regular languages; that is, while the indexing is considered to be "chosen", the languages inside the class and the domain D are more considered as "given". The results in the next section will establish various facts on the indexings which show that the indexings are not completely free, but some aspects of them are determined by the class which has to be represented by the automatic family. See [4, 5, 8, 9, 16, 17, 25, 26, 28] for more information on automatic structures.

2. The Size of Languages Inside a Family

Having the concept of an automatic family, one can use its indexing in order to introduce a measure for the size of the languages inside the given family.

Definition 2.1: Given an automatic family $\mathcal{L}$ and a language $R \in \mathcal{L}$, let $d_{\mathcal{L}}(R) = \min\{|i| : i \in I \wedge L_i = R\}$ be the size of R.

The next result shows that the size depends only up to an additive constant on the chosen automatic family; so enlarging the underlying family or just changing its indexing has not much impact on the size of a languages inside the family.

Proposition 2.2: *Let $\mathcal{L} = \{L_i : i \in I\}$ and $\mathcal{H} = \{H_j : j \in J\}$ be two automatic families. Then there is a constant c such that $d_{\mathcal{L}}(R)$ and $d_{\mathcal{H}}(R)$ differ by at most c, for all $R \in \mathcal{L} \cap \mathcal{H}$.*

Proof: The basic idea is to look at the set

> $O = \{conv(i,j) : i \in I$ and $j \in J$ and $L_i = H_j$ and i, j are the length-lexicographically least indices of their respective sets$\}$.

Note that $(i = i' \vee j = j') \Rightarrow (i = i' \wedge j = j')$ whenever $conv(i,j)$, $conv(i',j') \in O$. The length-lexicographic order $<_{ll}$ is automatic. Hence O is first-order definable from automatic relations:

$$\begin{aligned} conv(i,j) \in O \Leftrightarrow \quad & i \in I \wedge j \in J \wedge \forall x \in D\,[x \in L_i \Leftrightarrow x \in H_j] \\ & \wedge \forall i' \in I\,[[\forall y \in D\,[y \in L_{i'} \Leftrightarrow y \in L_i]] \Rightarrow i \leq_{ll} i'] \\ & \wedge \forall j' \in J\,[[\forall y \in D\,[y \in H_{j'} \Leftrightarrow y \in H_j]] \Rightarrow j \leq_{ll} j']. \end{aligned}$$

From this fact it follows, by a result of Khoussainov and Nerode [17], that the set O is regular.

Now one uses the following version of the pumping lemma: If R is a regular language then there is a constant c such that for all $uvw \in R$ with $|v| > c$ it also holds that $utw \in R$ for some string t which consists of up to c symbols taken from v.

This version of the pumping lemma is now applied to O. Let c be the corresponding constant. Given any $conv(i,j) \in O$, one takes u to be the prefix of length $\min\{|i|,|j|\}$ of $conv(i,j)$, v to be the suffix of length $\max\{|i|,|j|\} - \min\{|i|,|j|\}$ and w to be the empty string. Assume that $|v| > c$ and let t be formed by up to c characters from v as described in the pumping lemma above. Now $ut \in O$. There are two cases. If $|j| > |i| + c$ then there is a shorter $j' \in J$ with $conv(i,j') = ut$ and $conv(i,j') \in O$;

otherwise $|i| > |j| + c$ and there is a shorter $i' \in I$ with $conv(i', j) = ut$ and $conv(i', j) \in O$. So in either case, there is besides $conv(i, j)$ another pair, namely $conv(i', j)$ or $conv(i, j')$, in O; this pair coincides with $conv(i, j)$ in one but not in both coordinates in contradiction to the choice of O. Hence it cannot happen that $|v| > c$. Thus, the length of i and j differ by at most the constant c. □

This result is a bit parallel to the corresponding result in the field of Kolmogorov complexity [20] that the Kolmogorov complexity of an object depends only up to a constant on the underlying universal machine. The main difference is that here the measures $d_{\mathcal{L}}$ are only defined on a subfamily $\mathcal{L}$ of the regular languages and not on all of them; this invokes some problems and in the following it is investigated to which degree one can overcome these problems. Before doing this in the next sections, first a parallel to Kolmogorov complexity is pointed out: Boolean operations and images of sets under functions essentially have the complexity of the input sets.

Remark 2.3: Let an automatic family $\mathcal{L} = \{L_i : i \in I\}$ and an automatic predicate Φ mapping n inputs $L_{i_1}, L_{i_2}, \ldots, L_{i_n}$ to a new set $\Phi(L_{i_1}, L_{i_2}, \ldots, L_{i_n})$ be given. Then there is a new automatic family $\mathcal{H}$ such that for every $i_1, i_2, \ldots, i_n$ it holds that $\Phi(L_{i_1}, L_{i_2}, \ldots, L_{i_n})$ is contained in $\mathcal{H}$ and $d_{\mathcal{H}}(\Phi(L_{i_1}, L_{i_2}, \ldots, L_{i_n})) \leq c + \max\{d_{\mathcal{L}}(L_{i_1}), d_{\mathcal{L}}(L_{i_2}), \ldots, d_{\mathcal{L}}(L_{i_n})\}$, where c is a constant only depending on $\mathcal{L}$, $\mathcal{H}$ and Φ. Examples of such operators Φ are the Boolean operations like union, intersection and complementations as well as forming the range under an automatic function f: $\Phi(L) = \{f(x) : x \in L\}$.

3. Universal Complexity Measures

In the following, let A_R be the smallest deterministic finite automaton accepting R; A_R has to be complete, that is, for every state p and every symbol $a \in \Sigma$ there is exactly one state q such that A_R goes from p to q on input a. Let $d_{dfa}(R)$ denote the number of states of A_R and $d_{run}(R)$ denote the maximum n such that, for some input word x, A_R on x goes through n different states. Note that $d_{run}(R) \leq d_{dfa}(R)$. The next result establishes an inequality for the converse direction. This inequality witnesses that, for each n, there are only finitely many R with $d_{run}(R) = n$. Thus d_{run} satisfies some minimum requirement for measuring the size of a language adequately.

Proposition 3.1: *Let $R \subseteq \Sigma^*$ be a regular language. If Σ has at least two members then*

$$d_{dfa}(R) \le \frac{|\Sigma|^{d_{run}(R)} - 1}{|\Sigma| - 1};$$

otherwise $d_{dfa}(R) = d_{run}(R)$.

Proof: Recall that A_R is the minimal deterministic finite automaton recognising R. One can look at the finite tree T of all runs of A_R in which no state is visited twice. The height of this tree T is at most $d_{run}(R) - 1$. Furthermore, every state of A_R occurs in this tree T as it can be reached by a repetition-free run. In the case that Σ has exactly one element, T has $d_{run}(R)$ members and $d_{dfa}(R) = d_{run}(R)$. In the case that Σ has at least two members, the formula

$$d_{dfa}(R) \le |T| \le |\Sigma|^0 + |\Sigma|^1 + \ldots + |\Sigma|^{d_{run}(R)-1} = \frac{|\Sigma|^{d_{run}(R)} - 1}{|\Sigma| - 1}$$

provides an upper bound on the number of members of T and thus on the value $d_{dfa}(R)$. □

Remark 3.2: The exponential bound in the case of the alphabet Σ having at least two symbols looks large, but the gap cannot be made much smaller.

The proof of this fact and later results use the notion of the derivative: $L[x] = \{y : xy \in L\}$ is called the *derivative of L at x.* Note that $d_{dfa}(L)$ coincides with the number of distinct derivatives of L.

The exponential bound is now witnessed by the example family $L_0, L_1, L_2, \ldots$ where $L_n = \{xx : x \in \Sigma^n\}$. Then $d_{run}(L_n) = 2n + 2$ while $d_{dfa}(L_n) \ge |\Sigma|^n$, as for every $x \in \Sigma^n$ the derivative $L_n[x] = \{x\}$ is encoding x.

The following connections hold between the size based on automatic families and these two measures.

Theorem 3.3: *For every automatic family $\mathcal{L}$ there is a constant c such that $d_{dfa}(R) \le d_{\mathcal{L}}(R) \cdot c + 1$ for all $R \in \mathcal{L}$.*

Proof: Let Σ be the alphabet satisfying $R \subseteq \Sigma^*$ for all $R \in \mathcal{L}$ and let A be the automaton recognising the automatic class. That is, there is a regular set I of indices such that $\mathcal{L} = \{L_i : i \in I\}$ and a deterministic finite automaton A which accepts $conv(i, x)$ iff $i \in I$ and $x \in L_i$. Let c be the number of states of A. Let n be the length of the index i of some L_i. Now one constructs an automaton $Comb(A, i)$ which recognises the language L_i and which has at most $n \cdot c + 1$ states. $Comb(A, i)$ is constructed as follows.

- The alphabet of $Comb(A, i)$ is Σ.
- For $m < n$ let $X_m = \Sigma^m$ and for $m = n$ let $X_m = \bigcup_{k \geq n} \Sigma^k$.
- The set of states of $Comb(A, i)$ is the union of sets $Q_0, Q_1, \ldots, Q_n$, where Q_m consists of all pairs (q, m) with q being a state in A such that for some $x \in X_m$, A is in state q after reading the first $|x|$ symbols of $conv(i, x)$.
- For $m < n$, let there be a transition from (q, m) to $(p, m+1)$ on symbol a if there is a word $x \in X_{m+1}$ of length $m + 1$ such that A is in state q after reading first m symbols of $conv(i, x)$, A is in state p after reading first $m + 1$ symbols of $conv(i, x)$ and the last symbol of x is a. Furthermore, let there be a transition from (q, n) to (p, n) on symbol a if there is a word $x \in X_n$ such that A after reading $conv(i, x)$ is in state q and after reading $conv(i, xa)$ is in state p. There are no other transitions.
- Note that Q_0 contains only $(s, 0)$ where s is the starting symbol of A; $(s, 0)$ is then the starting state of the automaton $Comb(A, i)$.
- The accepting states of $Comb(A, i)$ are all states of the form (p, m) such that there is an $x \in X_m \cap L_i$ and $Comb(A, i)$, on input x, goes from state $(s, 0)$ to state (p, m).

Note that $|Q_0| = 1$ and, in general, $|Q_m| \leq c$. Hence $Comb(A, i)$ has at most $n \cdot c + 1$ states. Furthermore, one can show that $Comb(A, i)$, on input x, goes from state $(s, 0)$ to state (q, m) iff $x \in X_m$ and A, after reading first $|x|$ symbols of $conv(i, x)$, goes from starting state s to state q. Assume now that x, y are such that $Comb(A, i)$ is in the same state (q, m) after processing x, y. Clearly $x, y \in X_m$. If $m < n$, then after reading the first m symbols of $conv(i, x)$ and $conv(i, y)$, respectively, A is in the same state. Therefore A accepts $conv(i, x)$ iff A accepts $conv(i, y)$; hence $x \in L_i$ iff $y \in L_i$. Furthermore, if $m = n$, then after reading $conv(i, x)$ and $conv(i, y)$, A is in the same state q. Again $x \in L_i$ iff $y \in L_i$. It follows that $H_{q,m} = \{x \in \Sigma^* : Comb(A, i)$, on input x, goes from state $(s, 0)$ to state $(q, m)\}$ is either a subset of L_i or disjoint to L_i. So, by the definition of $Comb(A, i)$, the automaton $Comb(A, i)$ accepts the members of $H_{q,m}$ iff $H_{q,m} \subseteq L_i$ and rejects the members of $H_{q,m}$ iff $H_{q,m} \cap L_i = \emptyset$. It follows that $Comb(A, i)$ recognises the language L_i. The minimal automaton of L_i has at most as many states as $Comb(A, i)$. So $d_{dfa}(L_i) \leq |i| \cdot c + 1$ and therefore the theorem follows. □

Remark 3.4: The multiplicative constant c in the above theorem is indeed needed: If $\mathcal{L}$ is the class of all finite languages consisting of up to c strings, then each automaton accepting the language $L_n = \{0^d 1^n 0^d : d < c\}$ needs at least $c \cdot n$ states in order to memorise the number of 0s and then to count the number of 1s before comparing the number of 0s after the block of 1s with the memorised value. Also, one can easily verify that — up to an additive constant — $d_{\mathcal{L}}(L_n) = n$.

The next result is the main contribution of this paper. It shows that d_{run} is a measure which meets the expectation in at least one point: given any automatic class $\mathcal{L}$, the measures $d_{\mathcal{L}}$ and d_{run} coincide on $\mathcal{L}$ up to a constant.

Theorem 3.5: *For every automatic family $\mathcal{L}$ there is a constant c' such that, for all $R \in \mathcal{L}$, the values $d_{run}(R)$ and $d_{\mathcal{L}}(R)$ differ from each other by at most c'.*

Proof: Let $\mathcal{L}$ be the given automatic family and let c be the number of states of the minimal automaton A recognising the family. Let $R \in \mathcal{L}$.

Now it is shown that $d_{run}(R) \leq d_{\mathcal{L}}(R) + c$. For every index $i \in I$, one constructs the automaton $Comb(A, i)$ as described in Theorem 3.3. In any run, this automaton passes through at most $|i| + c$ states; namely for each $m < |i|$ through at most one state in Q_m and through at most c states in $Q_{|i|}$. On input x, the minimal automaton A_R goes through at most as many states as $Comb(A, i)$. Thus the bound obtained is also a bound for $d_{run}(R)$.

Now it is shown that $d_{\mathcal{L}}(R)$ is bounded by $d_{run}(R)$ plus a constant independent of R. For $i \in I$, let $L_i[x] = \{y : xy \in L_i\}$ be a derivative of the language L_i. Now, if $|x| \geq |i|$ then one can produce the following automaton B accepting the derivative $L_i[x]$:

- The set of states of B equals the set of states of A;
- The starting state of B is the state of A after having read the first $|x|$ symbols of $conv(i, x)$;
- The state transition of B from p to q on a symbol a occurs iff A goes on the one-symbol word $conv(\diamond, a)$ from p to q;
- The accepting states of B and A are the same.

Hence the language $L_i[x]$ is recognised by an automaton containing only c states. Let B_q be the automaton B with the starting state q. Now let n be the least positive natural number such that, for every $x \in \{0,1\}^n$, $L_i[x]$ is recognised by some automaton B_q. Note that the choice of q only depends on the state in which A is after having read first $|x|$ symbols of $conv(i, x)$. Without loss of generality, some of the B_q accept the empty string and some do not. Now let $t : \{1, 2, \ldots, c\} \to \{1, 2, \ldots, c\}$ code a finite function satisfying the following conditions:

- if there is $x \in \Sigma^n$ and A is in state b after processing the first n symbols of $conv(i, x)$ then $B_{t(b)}$ recognises $L_i[x]$;
- if there is $y \in \Sigma^*$ with $|y| < n$ such that A is in state b after processing the first n symbols of $conv(i, y)$ then $B_{t(b)}$ accepts the empty string iff $y \in L_i$.

Note that if there are $x, y \in \Sigma^*$ with $|x| = n \wedge |y| < n$ and A being in the same state after processing the first n symbols of $conv(i, x)$ and $conv(i, y)$, respectively, then $x \in L_i$ iff $y \in L_i$ iff the empty string is in $L_i[x]$. So above conditions do not contradict each other and the mapping t exists.

Using n and t, one can code L_i by an index j which is the convolution of t and the first n symbols of i. Note that there is an automatic function f with $f(i) = j$ as above; the reason is that n and t can be defined from i and the indexing of $\mathcal{L}$ using first-order formulas. Let $J = \{f(i) : i \in I\}$ and $H_{f(i)} = L_i$; the set J is regular and the family $\{H_j : j \in J\}$ is automatic.

Now let $i \in I$ and $j = f(i)$. It follows, using the pumping lemma, that every word in L_i of length $d_{run}(L_i)$ is of the form uvw such that $L_i[uv^{|i|}w] = L_i[uvw]$. Hence $L_i[uvw]$ is recognised by one of the automata B_q and so $|j| \leq d_{run}(L_i)$. Thus it holds, for all $R \in \mathcal{L}$, that $d_{run}(R) \geq d_{\mathcal{H}}(R)$. As $d_{\mathcal{L}}(R)$ and $d_{\mathcal{H}}(R)$ differ by at most a constant (Proposition 2.2), it follows that $d_{\mathcal{L}}(R) \leq d_{run}(R) + c''$, for some constant c''.

It follows from above analysis that $d_{run}(R)$ differs from $d_{\mathcal{L}}(R)$ only by a constant independent of R. □

Corollary 3.6: *For every automatic family $\mathcal{L}$ there is a constant c such that $d_{\mathcal{L}}(R) \leq d_{dfa}(R) + c$ for all $R \in \mathcal{L}$.*

The following result shows that — when restricted to an automatic family — the size of Boolean operations and images among the members of the family do not have a much larger size than the corresponding components. This situation is similar to the situation with respect to Kolmogorov complexity.

Theorem 3.7: *For every automatic family $\mathcal{L}$ over domain D and every automatic function f over domain D, there is a constant c such that, for all $L, H \in \mathcal{L}$, it holds that*

- $d_{run}(L \cup H) \leq \max\{d_{run}(L), d_{run}(H)\} + c$,
- $d_{run}(L \cap H) \leq \max\{d_{run}(L), d_{run}(H)\} + c$,

- $d_{run}(D - L) \leq d_{run}(L) + c$ *and*
- $d_{run}(\{f(x) : x \in L\}) \leq d_{run}(L) + c.$

However the first, second and fourth condition of this list do not hold without the restriction to an automatic family.

Proof: The main part of this result follows from the Proposition 2.2, Remark 2.3 and Theorem 3.5. So the rest of the proof is to show that these connections do not hold in general, except for the third condition which is just obtained by interchanging acceptance and rejection inside D.

Let $n \in \{1, 2, 3, \ldots\}$. For the first two conditions one considers the languages $(\Sigma^n)^*$ and $(\Sigma^{n+1})^*$. While $d_{run}((\Sigma^n)^*) = n$ and $d_{run}((\Sigma^{n+1})^*) = n+1$, it holds that $d_{run}((\Sigma^n)^* \cup (\Sigma^{n+1})^*) = n(n+1)$ and $d_{run}((\Sigma^n)^* \cap (\Sigma^{n+1})^*) = n(n+1)$. For the fourth condition, assume $\Sigma = \{0, 1, 2\}$ and consider the automatic function f which interchanges 1 and 2 at every second occurrence of one of these digits. So $f(001001001001) = 001002001002$ and $f(1212121221212121) = 1111111122222222$. Let $L = (0^n 1)^*$. Then $\{f(x) : x \in L\} = (0^n 1 0^n 2)^* + (0^n 1 0^n 2)^* \cdot 0^n 1$. While $d_{run}(L) = n + 2$ it holds that $d_{run}(\{f(x) : x \in L\}) = 2n + 3$. □

Remark 3.8: Besides d_{run}, one could also look at the following measure: $d_{rf}(R)$ is the largest number n such that there is an input $x \in \Sigma^*$ on which the minimal automaton A_R for R goes through exactly n states without repeating any of them.

Note that $d_{rf}(R) \leq d_{run}(R) \leq d_{dfa}(R)$ for every regular language R and the proof of Proposition 3.1 gives directly that

$$d_{dfa}(R) \leq \frac{|\Sigma|^{d_{rf}(R)} - 1}{|\Sigma| - 1}$$

for the case that Σ has at least two elements. Furthermore, Theorem 3.5 holds also with d_{rf} in place of d_{run}.

Although there are many parallel results for these two measures, d_{run} and d_{rf} are not identical. Consider the language $R = \{0^n, 1^n\}^*$ with $n \geq 2$. The minimal automaton A_R has the states reached by 0^m with $m < n$, 1^m with $m < n$ plus one rejecting state which is never left; as the initial state is double counted in this list, there are in total $2n$ states and $d_{dfa}(R) = 2n$. On the word $0^n 1^n 01$ all states are visited and therefore $d_{run}(R) = 2n$. However $d_{rf}(R) = n + 1$ and this maximum number is taken on the input $0^{n-1}1$.

4. Characterising Automatic Families

The central question of this section is: When is a class $\mathcal{L}$ of regular languages a subclass of some automatic family. The answer is that every language in the class must be representable by an automaton of a specific form.

Theorem 4.1: *A class $\mathcal{L}$ is a subclass of an automatic family iff there is a constant c such that every $R \in \mathcal{L}$ is accepted by a deterministic finite automaton whose states can be partitioned into sets $Q_0, Q_1, Q_2, \ldots, Q_n$ satisfying the following conditions: each set Q_m has up to c states; Q_0 consists exactly of the starting state; if there is a transition from a state p to a state q, then there are r, r' with $p \in Q_r$, $q \in Q_{r'}$ and $r' = \min\{r+1, n\}$.*

Proof: Given an automatic class $\mathcal{L}$ with an automaton A recognising the class, one can construct, for each $i \in I$, the automaton $Comb(A, i)$ to recognise L_i as in Theorem 3.3; it is easy to see that the automaton is of the above form.

For the converse direction, consider the class $\mathcal{H}$ of all languages which are accepted by an automaton of the above form with a given fixed constant c. The indices j of $\mathcal{H}$ consist of symbols coding $Q_0, Q_1, \ldots, Q_n$, respectively. For each Q_m it is coded which of the states of Q_m (numbered as $1, 2, \ldots, c$) are accepting and what transitions are there from Q_m to $Q_{\min\{m+1,n\}}$ based on various inputs from Σ. Without loss of generality state 1 from Q_0 is the starting state and that does not need to be coded. Now $J = \Gamma^*$ where

$$\Gamma = \{\text{reject}, \text{accept}\}^c \times \{1, 2, \ldots, c\}^{|\{1,2,\ldots,c\} \times \Sigma|},$$

that is, where each symbol in Γ codes the acceptance of the c states in Q_m plus the transition table to $Q_{\min\{m+1,n\}}$. Without loss of generality, the empty string just codes the empty language. This convention permits to avoid a domain check for the index j.

As a finite automaton is the same as an algorithm working from the left to the right through the word with constant memory, the algorithm is now given more explicitly than it would be in the case of an automaton. It runs in stages and it memorises information which can be stored in constantly many bits; note that the number of these bits depends on the value of c but not of the value of n. For easier readability, this information is memorised in variables which are initialised in the first two steps. The algorithm reads in steps (2) and (4) the symbols unless the end of the corresponding inputs (j and x) is reached in which case the last value is not overwritten. If j is empty, then the algorithm rejects all x.

(1) Variables: status $\in$ {reject,accept}; state $\in \{1, 2, \ldots, c\}$; $a \in \Sigma$ (current input symbol to be processed); b (table of current Q_m).
Let state = 1 and b be a code such that 1 is a rejecting state and all transitions are from 1 to 1.
(2) If there is some symbol of the code of j to be read
then read b
else let b unchanged.
(3) If the current value of state according to b is an accepting state
then let status = accept
else let status = reject.
(4) If there is some symbol of the input word to be read
then read a
else accept/reject according to status and terminate.
(5) Decode from b the new value of state in dependence of a and of the current value of state.
(6) Go to (2).

The verification is left to the reader, as it is straightforward but lengthy. Note that $\mathcal{L}$ is automatic iff the set $\{j \in J : H_j \in \mathcal{L}\}$ is regular. □

The characterisation from Theorem 4.1 could be put into a more general form. Recall that $R[x]$ is the set $\{y \in \Sigma^* : xy \in R\}$.

Theorem 4.2: *A class $\mathcal{L}$ is a subclass of an automatic family iff there is a constant c such that $R \in \mathcal{L}$ iff there is an n such that:*

- *For every m there are at most c different derivatives $R[x]$ with $x \in \Sigma^m$;*
- *There are at most c different derivatives $R[x]$ with $x \in \Sigma^* \wedge |x| \geq n$.*

This result has an interesting corollary for the case of the unary alphabet.

Corollary 4.3: *If $\Sigma = \{0\}$ then $\mathcal{L}$ is contained in an automatic family iff there is a finite class $\mathcal{F}$ of regular languages such that every language in $\mathcal{L}$ is equal to $H \cup (0^m \cdot L)$ for some m, some $L \in \mathcal{F}$ and some $H \subseteq \{0^\ell : \ell \leq m\}$.*

In learning theory, an important and well-studied family is that of the pattern languages [2, 18, 27]. Here a pattern is a string consisting of constants and variables, where the language generated by the pattern is the set of all those words which are obtained by replacing variables by strings. If the strings replacing the variables are permitted to be empty, then these languages are called "erasing pattern languages"; if these strings are not permitted to be empty, then these languages are called "non-erasing pattern languages". The learnability of pattern languages has been extensively studied and while there is an explanatory learner for the class of non-erasing pattern languages [2], such a learner does not exist in the case of the erasing pattern languages [24]. In particular, the class of "regular pattern languages" is quite important and one might ask what the automatic counterpart of it is. Here Shinohara [27] defined that a pattern is called a *regular pattern* iff it contains every variable at most once. So if $\Sigma = \{0, 1, 2\}$ and the pattern is $i = 01121x121y112z$ then the language generated by this pattern is given by the regular expression $01121 \cdot (0+1+2)^* \cdot 121 \cdot (0+1+2)^* \cdot 112 \cdot (0+1+2)^*$; in the non-erasing case, one would have to replace "$(0+1+2)^*$" by "$(0+1+2) \cdot (0+1+2)^*$" as every variable represents at least one letter.

Theorem 4.4: *Let $\mathcal{L}$ be a class of erasing pattern languages, each generated by a regular pattern. The class $\mathcal{L}$ is contained in an automatic family iff there is a constant c such that, in every pattern of a language in the family, there are at most c constants after the occurrence of the first variable.*

Proof: In the case that Σ has only one element, say 0, variables and constants commute and every pattern language in the class is generated by a pattern where the constants come first and the variables come last. One can without loss of generality then assume that there is at most one variable. Hence over the unary alphabet, the class of all erasing regular pattern languages is contained in the automatic family of the languages generated by one of the patterns in $\{x, 0x, 00x, 000x, \ldots\} \cup \{\epsilon, 0, 00, 000, \ldots\}$. So assume the case that Σ has at least two elements.

Given the constant c, it is first shown that there is an automatic family containing all the erasing regular pattern languages with up to c constants after the first occurrence of a variable. Note that there are only finitely many regular patterns which start with a variable and contain up to c constants; the reason is that a double variable xy has the same effect as a single variable in the case that variables can take the empty word and are not repeated. Now let Γ be the set of all these patterns and assume that Γ is coded such that it is disjoint to Σ. One chooses as an indexing the set $\Sigma^*\Gamma$. If L_g is the language generated by $g \in \Gamma$ then extend this definition to $L_{ig} = i \cdot L_g$ for $i \in \Sigma^*$. The family of all L_{ig} is automatic as the finite automaton recognising $\{conv(ig, x) : x \in L_{ig}\}$ accepts $conv(ig, x)$ iff $x = iy$ for some y and $y \in L_g$; note that the latter can be checked as there are only finitely many languages of the form L_g with $g \in \Gamma$ — thus, one can combine the corresponding finite automata to get an automaton for checking whether $y \in L_g$.

For the converse direction, assume that an automatic family $\{L_i : i \in I\}$ of erasing regular pattern languages is given. As shown in the proof of Theorem 3.5, there is a constant c such that, for every $i \in I$, there are at most c different derivatives $L_i[x]$ with $|x| \geq |i|$. Now assume that $i \in I$ is the index of the language generated by a pattern of the form $ux_0a_1x_1 \ldots a_nx_n$ where $u \in \Sigma^*$, $a_1, \ldots, a_n \in \Sigma$, x_0 is a variable and $x_1, \ldots, x_n$ are each either a variable or the empty string. Furthermore, let $b \in \Sigma - \{a_1\}$. Now choose k larger than the length of i. Then there are n different derivatives $L_i[ub^k a_1 \ldots a_m]$ correspondingly with the shortest word $a_{m+1} \ldots a_n$, where $m \in \{1, 2, \ldots, n\}$. Thus, as $|ub^k| \geq |i|$ the inequality $n \leq c$ holds and therefore the pattern has at most c constants after the occurrence of the variable x_0. □

By using essentially the same proof, one can obtain the counterpart of this result for non-erasing pattern languages. As here the variables have at least the length 1, the patterns x and yz do not generate the same language; hence one has to bound the number of variables as well.

Corollary 4.5: *Let $\mathcal{L}$ be a class of non-erasing pattern languages, each generated by a regular pattern. The class $\mathcal{L}$ is contained in an automatic family iff there is a constant c such that every language in $\mathcal{L}$ can be generated by a pattern which has at most c variables and constants after the first block of variables. That is, if the pattern contains a constant a after some variable x, then there are at most c variables and constants after the above xa.*

A direct application of this result is the following: The family of languages generated by the patterns $\{x, 0x, 00x, 000x, 0000x, 00000x, \ldots\}$ is automatic while the family of languages generated by the patterns $\{x_00, x_00x_1, x_00x_1x_2, x_00x_1x_2x_3, \ldots\}$ is, in the case that $|\Sigma| \geq 2$, not automatic.

Note that every regular pattern generates a regular language [27]. Reidenbach [23] discussed the converse direction and showed that certain non-erasing patterns generate regular languages although the patterns themselves are not regular; he furthermore noted that in the non-erasing case, a language is either generated only by regular patterns or only by nonregular patterns. Jain, Ong and Stephan [13] show that the converse direction also fails for erasing pattern languages and alphabet sizes up to 3.

Remark 4.6: Reidenbach [23] provided examples of nonerasing pattern languages which are regular but are not generated by a regular pattern. An example is given by the pattern $xyxz$ which generates the regular language

$\bigcup_{a,b,c\in\Sigma} ab\Sigma^* ac\Sigma^*$ and which, for alphabets of size 2 or more, cannot be generated by a regular pattern.

Jain, Ong and Stephan [13] considered the corresponding question for erasing pattern languages. If Σ has at least four symbols, then every erasing pattern language which is regular is generated by a pattern which does not have repetitions of the variables. However for alphabets of size 1, 2 and 3 there are also erasing pattern languages which are regular sets but need repetitions of variables to be generated. For $\Sigma = \{0, 1\}$ the pattern $x_1x_2x_31x_2x_4x_4x_51x_6x_5x_7$ generates the language $(0+1)^* \cdot 1 \cdot (0+1)^* \cdot 1 \cdot (0+1)^* - 10 \cdot (00)^* \cdot 1$, which is not generated by any regular pattern.

5. Applications of Automatic Families in Learning Theory

Automatic families are a special case of indexed families [1, 19] which are a widely studied subject in inductive inference. Here an indexed family $\{L_i : i \in I\}$ is uniformly recursive, that is, the domain D, the set of admissible indices I and the mapping $i, x \mapsto L_i(x)$ are all recognisable by Turing machines. So automatic families are just the restriction obtained when replacing Turing machines by finite automata with possibly several inputs. Gold [7] formalised the notion of learning and introduced the following notion of *explanatory learning*: A family $\{L_i : i \in I\}$ is explanatorily learnable iff there is a recursive learner which reads more and more data about the language R to be learnt from a text T and outputs a sequence $e_0, e_1, e_2, \ldots$ of indices which syntactically converges to one index $i \in I$ with $R = L_i$. Here a text is an infinite sequence of words containing every member of the set to be learnt but not any nonmember; the words in the text can be in arbitrary and adversary order and repetitions are permitted. The interested reader might find more background information on learning theory in the standard textbooks of inductive inference [14, 22]. Angluin [1] gave a criterion on the learnability of an indexed family which — in the case of automatic families — can be simplified to the following one [11].

Theorem 5.1: *An automatic family $\mathcal{L} = \{L_i : i \in I\}$ is explanatorily learnable iff there exists a constant c such that for all $i, j \in I$ the equality $L_i = L_j$ holds whenever $\{x \in L_i : |x| \leq d_{\mathcal{L}}(L_i) + c\} \subseteq L_j \subseteq L_i$.*

In other words, one can build a learner which — when learning R — always conjectures the least i such that all data in L_i shorter than $|i| + c$ have already been observed but no data outside L_i.

Based on this observation, Jain, Luo and Stephan [11] formulated the notion of an automatic learner which is less powerful than a recursive learner.

An automatic learner has a long term memory which stores all relevant information about the data observed; this long term memory is a string like any input word, though it might be over a larger alphabet. The learner is then given by an update function

> F: (old long term memory, current data item) $\mapsto$ (new long term memory, new hypothesis).

Starting from an initial value for its long term memory, the learner reads in each round a current datum and updates its memory and the hypothesis according to F. The update function F has to be automatic. Jain, Luo and Stephan [11] showed that not every learnable automatic class can also be learnt by an automatic learner. Indeed even quite simple classes like the one given by $L_i = \Sigma^* - \{i\}$ (where $D = I = \Sigma^*$) is learnable only if the alphabet consists of one symbol; in the case of an alphabet with at least two symbols this class is no longer learnable by an automatic learner, as the automatic learner cannot memorise enough information about the data observed. Special cases considered were those where the long term memory cannot be longer than the longest datum observed so far, where the long term memory is the last hypothesis conjectured (iterative learning) and where the long term memory consists of up to c data items observed previously (bounded example memory); the ability for an automatic learner to learn depends heavily on the nature of such restrictions imposed. Ong [21] investigated the learnability of automatic families of pattern languages and related classes.

Jain, Martin and Stephan [12] considered the setting of robust learning [6, 15, 29] and asked when every translation of an automatic family $\mathcal{L}$ is learnable, where a translation is given by a first-order definable operator which preserves inclusions among all languages as well as non-inclusions among languages from the class. An example is $\Phi(L) = \{x : \exists y \in L\, [y \neq x]\}$ and the underlying family $\mathcal{L}$ contains $\emptyset$, every singleton $\{x\}$ and the full set D. It is easy to see that $\mathcal{L}$ is learnable. However the given translation of $\mathcal{L}$ is not learnable, as the learner would have to converge to $\Phi(D)$ after having seen finitely many data items and then the input language could be $\Phi(\{y\}) = D - \{y\}$ for some y which has not yet been observed in the input. General characterisations were given for the question of the following type: "For which classes are all translations learnable under a given criterion?" Besides the standard criteria from inductive inference, the paper also looked at query learning. Here a learner can ask a teacher questions, in a given query language; for example the learner can ask whether the language

L_i is a superset of the language R to be learnt (if the query language allows superset queries). The learner asks finitely many queries and has then to conjecture the correct index of the input language. Angluin [3] started the investigation of the learnability of regular languages from queries. She showed that the class of all regular languages can be learnt using membership and equivalence queries in polynomial time where, when learning R, the time bound depends on $d_{dfa}(R)$ and the largest counter example observed. The following result of Jain, Martin and Stephan [12] links explanatory learning, query learning and robustness.

Theorem 5.2: *The following conditions are equivalent for an automatic family $\mathcal{L} = \{L_i : i \in I\}$.*
(a) For every $i \in I$ there is a bound b such that, for all $j \in I$ with $L_j \subset L_i$, there is a $k \in I$ with $k \leq_{ll} b$, $L_j \subseteq L_k$ and $L_i \not\subseteq L_k$.
(b) Every translation of $\mathcal{L}$ is explanatorily learnable.
(c) Every translation of $\mathcal{L}$ can be learnt using superset queries and membership queries.
(d) $\mathcal{L}$ can be learnt using superset queries.

Note that in condition (d) it does not matter whether one learns only $\mathcal{L}$ using superset queries or every translation of $\mathcal{L}$ using superset queries. The reason for this is that translations do not change the inclusion structure of the automatic family. This result shows that there are also connections to robust notions of query learning.

Jain, Luo, Semukhin and Stephan [10] investigated the question on what can be said on the learnability of uncountable families. For this they use ω-automatic families, where the indices of the sets are ω-words and where a nondeterministic Büchi-automaton checks whether a finite word x belongs to the language defined by an ω-word. Also here the Büchi automaton is fed with the convolution of the input word and the ω-word serving as the index of the language. As there are uncountably many indices, it is no longer possible for the learner to come up with the correct index after finite time; therefore the model is adjusted to a verification game. So the learner reads in parallel a text consisting of all the words in the language and an ω-index; the output is a sequence of Büchi automata which converges to one fixed automaton. Then this automaton has to accept the given ω-index iff it is an index for the language observed. Also in this setting, learnability is equivalent to Angluin's tell-tale condition [1]. However it is necessary for this result that the learner has the right to choose the indexing; otherwise the criterion is more restrictive. Also further other criteria are transferred to this model and it is shown that one can abstain from adjusting the indexing if one requires vacillatory learning, where the learner in the limit oscillates between finitely many Büchi automata, and each of them accepts the given ω-index iff it is an index for the language to be learnt.

References

1. Dana Angluin. Inductive inference of formal languages from positive data. *Information and Control* 45:117–135, 1980.
2. Dana Angluin. Finding patterns common to a set of strings. *Journal of Computer and System Sciences*, 21:46–62, 1980.
3. Dana Angluin. Learning regular sets from queries and counterexamples. *Information and Computation*, 75:87–106, 1987.
4. Achim Blumensath. *Automatic structures.* Diploma thesis, RWTH Aachen, 1999.
5. Achim Blumensath and Erich Grädel. Automatic structures. *15th Annual IEEE Symposium on Logic in Computer Science* (LICS), pages 51–62, IEEE Computer Society, 2000.
6. Mark Fulk. Robust separations in inductive inference. *Proceedings of the 31st Annual Symposium on Foundations of Computer Science* (FOCS), pages 405–410, St. Louis, Missouri, 1990.
7. E. Mark Gold. Language identification in the limit. *Information and Control* 10:447–474, 1967.
8. Bernard R. Hodgson. *Théories décidables par automate fini.* Ph.D. thesis, University of Montréal, 1976.
9. Bernard R. Hodgson. Décidabilité par automate fini. *Annales des sciences mathématiques du Québec*, 7(1):39–57, 1983.
10. Sanjay Jain, Qinglong Luo, Pavel Semukhin and Frank Stephan. Uncountable automatic classes and learning. *Algorithmic Learning Theory*, Twentieth International Conference, ALT 2009, Porto, Portugal, October 3-5, 2009. Proceedings. Springer LNAI 5809:293–307, 2009. Technical Report TRB1/09, School of Computing, National University of Singapore, 2009.
11. Sanjay Jain, Qinglong Luo and Frank Stephan. Learnability of automatic classes. *Language and Automata Theory and Applications*, 4th International Conference, LATA 2010, Trier, Germany, May 24-28, 2010. Proceedings. Springer LNAI 6031:321-332, 2010.
12. Sanjay Jain, Eric Martin and Frank Stephan. *Robust learning of automatic classes of languages.* Manuscript, 2010.
13. Sanjay Jain, Yuh Shin Ong and Frank Stephan. *Regular patterns, regular languages and context-free languages. Information Processing Letters* 110:1114–1119, 2010.
14. Sanjay Jain, Daniel N. Osherson, James S. Royer and Arun Sharma. *Systems That Learn.* MIT Press, 2nd Edition, 1999.
15. Sanjay Jain and Frank Stephan. *A tour of robust learning. Computability and Models. Perspectives East and West.* Edited by S. Barry Cooper and Sergei

S. Goncharov. Kluwer Academic / Plenum Publishers, University Series in Mathematics, pages 215–247, 2003.

16. Bakhadyr Khoussainov and Mia Minnes. Three lectures on automatic structures. *Proceedings of Logic Colloquium 2007. Lecture Notes in Logic*, 35:132–176, 2010.
17. Bakhadyr Khoussainov and Anil Nerode. Automatic presentations of structures. *Logical and Computational Complexity*, (International Workshop LCC 1994). Springer LNCS 960:367–392, 1995.
18. Steffen Lange and Rolf Wiehagen. Polynomial time inference of arbitrary pattern languages. *New Generation Computing*, 8:361–370, 1991.
19. Steffen Lange and Thomas Zeugmann. Language learning in dependence on the space of hypotheses. *Proceedings of the Sixth Annual Conference on Computational Learning Theory* (COLT), pages 127–136. ACM Press, 1993.
20. Ming Li and Paul Vitányi. *An Introduction to Kolmogorov Complexity and Its Applications.* Third Edition, Springer, 2008.
21. Yuh Shin Ong. *Automatic Structures and Learning.* Final Year Project, School of Computing, National University of Singapore, 2010.
22. Daniel Osherson, Michael Stob and Scott Weinstein. *Systems That Learn, An Introduction to Learning Theory for Cognitive and Computer Scientists.* Bradford — The MIT Press, Cambridge, Massachusetts, 1986.
23. Daniel Reidenbach. *The Ambiguity of Morphisms in Free Monoids and its Impact on Algorithmic Properties of Pattern Languages.* Logos Verlag, Berlin, 2006.
24. Daniel Reidenbach. A non-learnable class of E-pattern languages. *Theoretical Computer Science*, 350:91–102, 2006.
25. Sasha Rubin. *Automatic Structures.* Ph.D. Thesis, University of Auckland, 2004.
26. Sasha Rubin. Automata presenting structures: a survey of the finite string case. *The Bulletin of Symbolic Logic*, 14:169–209, 2008.
27. Takeshi Shinohara. Polynomial time inference of extended regular pattern languages. *RIMS Symposia on Software Science and Engineering, Kyoto, Japan, Proceedings.* Springer LNCS 147:115–127, 1982.
28. Wai Yean Tan. *Automatic Structures.* Honours Year Thesis, Department of Mathematics, National University of Singapore, 2008.
29. Thomas Zeugmann. On Bārzdiņš' Conjecture. *Proceedings of the International Workshop on Analogical and Inductive Inference* (AII'86), Springer LNCS 265:220–227, 1986.

CAPPABLE CEA SETS AND RAMSEY'S THEOREM

Asher M. Kach
Department of Mathematics, University of Connecticut
Storrs, CT 06269-3009
kach@math.uconn.edu

Manuel Lerman
Department of Mathematics, University of Connecticut
Storrs, CT 06269-3009
lerman@math.uconn.edu

Reed Solomon
Department of Mathematics, University of Connecticut
Storrs, CT 06269-3009
solomon@math.uconn.edu

We begin a search for degree-theoretic properties that might be used to separate Ramsey's Theorem for pairs from its stable version in the Reverse Mathematical sense. This paper introduces the notion of *c-cappability* and shows that this property cannot be used to obtain such a separation when combined with 2-CEA-ness.

2000 *Mathematics Subject Classification*. Primary: 03C57; Secondary: 03D45, 06A06.

1. Introduction

Let $\mathbb{N}$ denote the natural numbers. Fix $X \subset \mathbb{N}$, and let $[X]^2 = \{Y \subseteq X : |Y| = 2\}$. A *2-coloring* C of $[\mathbb{N}]^2$ is a function from $[\mathbb{N}]^2$ into $\{0, 1\}$; such a 2-coloring is said to be *stable* if for each $x \in \mathbb{N}$ there exists a $y \in \mathbb{N}$ and a $c \in \{0, 1\}$ such that $C(x, z) = c$ for all $z > y$. A set $H \subset \mathbb{N}$ is *homogeneous* for C if C is constant on $[H]^2$. Ramsey's Theorem for pairs, RT^2_2, states that every 2-coloring of $[\mathbb{N}]^2$ has an infinite homogeneous set, and Stable Ramsey's Theorem for pairs, SRT^2_2, is the analogous statement for stable 2-colorings of $[\mathbb{N}]^2$. An excellent summary of the reverse mathematical results dealing with Ramsey's Theorem can be found in [7].

This paper was motivated by an attempt to separate RT^2_2 from SRT^2_2, in the sense of Reverse Mathematics. One normally tries to achieve such a separation by building an ideal $\mathbf{I}$ of degrees such that every stable 2-coloring of $[\mathbb{N}]^2$ with degree in $\mathbf{I}$ has a homogeneous set in $\mathbf{I}$, but there is a (non-stable) 2-coloring of $[\mathbb{N}]^2$ with degree in $\mathbf{I}$ that does not have a homogeneous set in $\mathbf{I}$. Frequently, $\mathbf{I}$ is defined within a set of degrees that is closed under relativization such as the low$_n$ degrees. (Note that while the low$_n$ degrees are closed downward and under relativization, they are not closed under join and hence do not form an ideal.) For example, separations from WKL_0 are often obtained using the fact that there is an ideal $\mathbf{I}$ contained within the low degree such that the ω-model with second order part $\{X \mid \deg(X) \in \mathbf{I}\}$ is a model of WKL_0. Such an ideal suffices to show that WKL_0 cannot prove either RT^2_2 or SRT^2_2. (By results of Hirst, this separation can also be obtained by considering levels of induction in non-ω-models.)

It is not clear whether such ideals within the low or low$_2$ degrees can be used to separate SRT^2_2 and RT^2_2. With respect to the low$_2$ degrees, Cholak, Jockusch and Slaman [3] proved every computable instance of RT^2_2 (and hence also every computable instance of SRT^2_2) has a low$_2$ solution. With respect to the low degrees, while every computable instance of SRT^2_2 must have a Δ^0_2 solution (unlike computable instances of RT^2_2), Downey, Hirschfeldt, Lempp and Solomon [6] constructed a computable instance of SRT^2_2 which has no low solution. (Downey and Hirschfeldt announced recently that they can improve the result by replacing *low* with *low*$_2$.)

The aim of this paper is to begin a search for a smallness property of degrees which can be combined with low$_2$-ness to build a separating ideal. By *smallness*, we mean a property that is *low for Reverse Mathematics*; a degree is of this type if no degree $\geq \mathbf{0}'$ has that property. The property of degrees we consider is *c-cappability*; a degree $\mathbf{b}$ is said to be *c-cappable* if $\mathbf{b} \neq \mathbf{0}$ and there is a non-zero c.e. degree $\mathbf{a}$ and a CEA degree $\mathbf{d}$ such that $\mathbf{b} \leq \mathbf{d}$ and $\mathbf{a}$ and $\mathbf{d}$ form a minimal pair, i.e., the only degree below both $\mathbf{a}$ and $\mathbf{d}$ is $\mathbf{0}$. (See [8] for the definition of the REA sets and degrees, which have been renamed CEA.) We will show that c-cappability cannot be used to achieve such a separation if we require that the separating degree also be 2-CEA.

When considering lowness properties in the context of Reverse Mathematics, we mean a property not possessed by the degree $\mathbf{0}'$. It would be even more desirable if $\mathbf{0}'$ is not in the set generated by closing under finitely many applications of relativization. The low$_n$ degrees for each n are examples of sets of degrees with such properties. In searching for other possible properties to consider, we were led, by analogy, to look at the c-cappable degrees. Ambos-Spies, Jockusch, Shore and Soare [1] showed that the c-cappable c.e. degrees (i.e., the cappable c.e. degrees) form a proper ideal in the c.e. degrees, so are low in the sense just mentioned. As the CEA degrees are the closest analog found within the arithmetical degrees to the way the c.e. degrees sit below $\mathbf{0}'$, we hoped that the analogy would extend to show that the c-cappable degrees form an ideal within the arithmetical degrees which does not contain $\mathbf{0}'$. (Note that $\mathbf{0}'$ is not c-cappable.) Futhermore, we hoped that the same would be true under finitely many applications of relativization of c-cappability. The latter hope is refuted by a result of Ambos-Spies, Lempp and Lerman [2] which states that a finite lattice can be embedded into the c.e. degrees preserving least element 0 and greatest element 1 if and only if it can be partitioned into two parts; the first downward closed, containing the ideal generated by the cappable elements of the lattice, and having a greatest element a which is different from the greatest element of the lattice; and the second upward closed and containing a smallest element b. Figure 1 exhibits a finite lattice with this property in which $a \vee b = 1$, and so shows that $\mathbf{0}'$ lies in the ideal generated by the elements that are c-cappable relative to a c-cappable degree.

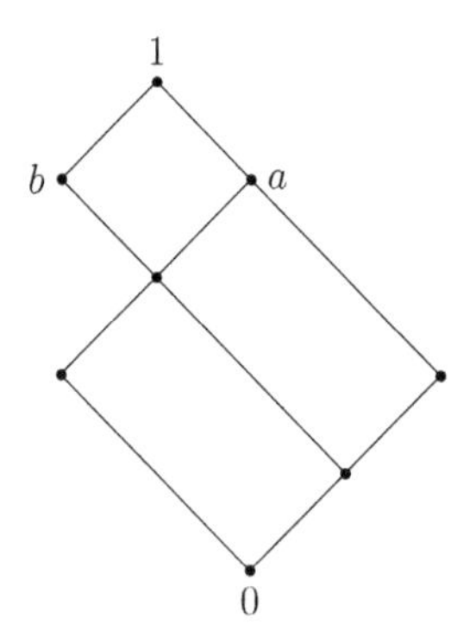

Fig. 1. Lattice example

We next note that Ding and Qian [4] showed that there is a non-zero c.e. degree $\mathbf{a}$ and a non-zero d-c.e. degree $\mathbf{b}$ whose meet is $\mathbf{0}$ and whose join is $\mathbf{0}'$. As Lachlan has shown that every d-c.e. degree is 2-CEA, it follows that the ideal generated by the c-cappable degrees contains $\mathbf{0}'$.

The failure of the closure under relativization of c-cappability and the fact that $\mathbf{0}'$ lies in the ideal generated by the c-cappable degrees still does not preclude the possibility of building an ideal of c-cappable degrees avoiding $\mathbf{0}'$; for example, such ideals exist within the low$_2$ degrees. Thus it still makes sense to ask whether every computable stable coloring of pairs has a homogeneous set of c-cappable degree. Our Theorem 10 shows that this, unfortunately, is not the case if we also require that the degree of the separating set be computable from a 2-CEA degree.

Our notation essentially follows that of [9]. We define the *use* of a computation to be the largest number whose membership in or out of the oracle is used in the computation, and let $B \upharpoonright u$ denote the string $\sigma \in \{0,1\}^{<\omega}$ of length $u+1$ such that $\sigma \subset B$. (Inclusion for strings is interpreted as extension, and sets are identified with their characteristic functions.)

2. SRT^2_2 and c-cappability

When we embarked on the research presented in this paper, we had hoped to show at least that every stable 2-coloring of pairs had a c-cappable 2-CEA homogeneous set. The main theorem of this section shows that this is not the case. Our proof relies on properties of enumerations of n-CEA sets, in particular, the case for $n = 2$. Our analysis is a generalization of what happens in the c.e. case, so we begin by reviewing it.

To simplify our definitions in this section, it is convenient to assume that all the enumerated sets are infinite. Therefore, rather than enumerating the standard sets W_e (possibly relative to some oracle), we enumerate the sets $W_e \oplus \mathbb{N}$. We modify the usual enumeration procedure for W_e by specifying that for all s, $2s+1$ enters $W_e \oplus \mathbb{N}$ at stage $s+1$ (with no oracle information in the cases when we are enumerating relative to an oracle). Thus we obtain the property that at every stage at least one number (and possibly more) is enumerated without any nonuniform worries about finite sets. Because we are only concerned with degree properties, this convention is permissible. We also retain the standard property that no numbers are enumerated at stage 0.

Throughout this section, we use $\subseteq$ to denote the subset relation between sets and $\preceq$ to denote the initial segment relation between strings. When appropriate, we view a set as an infinite string given by its characteristic function. The subscript 1 in the following definitions indicates that we are dealing with 1-CEA (that is, c.e.) sets. The subscript plays no role at this point, but will be relevant later when we extend up the n-CEA hierarchy.

Definition 1. Fix a c.e. set B_1 with an enumeration $\{B_1^s \colon s \in \mathbb{N}\}$, subject to the conventions above. (The definitions to follow depend on this enumeration, but we suppress this dependence in the notation, considering the enumeration fixed.) A stage t is a *1-nondeficiency stage* for B_1 if either $t = 0$ or $\min\{B_1^t - B_1^{t-1}\} < \min\{B_1 - B_1^t\}$. Similarly, for a stage s, we say that $t < s$ is a *$\langle 1, s\rangle$-nondeficiency stage* for B_1 if either $t = 0$ or $\min\{B_1^t - B_1^{t-1}\} < \min\{B_1^s - B_1^t\}$.

We use the nondeficiency stages to modify the approximation to B_1 as follows.

Definition 2. The *1-nondeficiency approximation* $\{\beta_1^s : s \in \mathbb{N}\}$ to B_1 is defined as follows. For $s = 0$, set $\beta_1^0 = \emptyset$. For $s > 0$, let β_1^s be such that $\beta_1^s \preceq B_1^s$ with length $\min\{B_1^s - B_1^{s-1}\}$. β_1^s is called the *1-nondeficiency oracle* for B_1 at stage s. We say that t is *1-correct* if $\beta_1^t \preceq B_1$ and that t is *$\langle 1, s\rangle$-correct* if $t < s$ and $\beta_1^t \preceq \beta_1^s$.

Notice that the nondeficiency approximation is really a Δ_2^0 approximation in the sense that when the length of β_1^s shrinks, the approximation temporarily removes numbers which are elements of B_1. The following facts are easy consequences of these definitions for any c.e. set B_1 with approximations as above.

Lemma 1. *If $0 = s_0 < s_1 < \cdots$ are the 1-nondeficiency stages for B_1, then $\beta_1^{s_0} \preceq \beta_1^{s_1} \preceq \cdots$. Furthermore, for any fixed s, if $0 = s_0 < s_1 < \cdots s_k$ are the $\langle 1, s\rangle$-nondeficiency stages for B_1 then $s_k < s$ and $\beta_1^{s_0} \preceq \beta_1^{s_1} \preceq \cdots \preceq \beta_1^{s_k} \preceq \beta_1^s$.*

Lemma 2. *A stage t is a 1-nondeficiency stage for B_1 if and only if t is 1-correct. For $t < s$, t is a $\langle 1, s\rangle$-nondeficiency stage for B_1 if and only if t is $\langle 1, s\rangle$-correct.*

Lemma 3. *For any $t < s$, if t is not $\langle 1, s\rangle$-correct, then t is not $\langle 1, r\rangle$-correct for any $r \geq s$ and t is not 1-correct.*

We generalize this concept to 2-CEA sets and then through the CEA hierarchy. We separate the 2-CEA case because it gives the general picture and we will need a special property of the 2-CEA level of the hierarchy for the main theorem of this section.

Consider a set B_2 which is c.e. relative to a c.e. set B_1. By convention, we assume that B_2 has the form $B_2 = W_e^{B_1} \oplus \mathbb{N}$ for some e and that the enumeration of B_2 from B_1 has the property that $2s + 1$ is enumerated into B_2 at stage $s + 1$ using no oracle information. We fix a computable approximation B_2^s to B_2 by $B_2^0 = \emptyset$ and for $s > 0$, B_2^s is the set of numbers enumerated using the 1-nondeficiency oracle β_1^s. Notice that this approximation has the property that $B_2^{s+1} - B_2^s \neq \emptyset$ for all s.

Definition 3. Let B_2 be c.e. relative to a c.e. set B_1 with a computable approximation B_2^s as described above. Let $0 = s_0 < s_1 < \cdots$ be the sequence of 1-nondeficiency stages for B_1. t is a *2-nondeficiency stage* for B_2 if $t = 0$ or $t = s_i$ for some $i > 0$ and $\min\{B_2^{s_i} - B_2^{s_{i-1}}\} < \min\{B_2 - B_2^{s_i}\}$. For $t < s$, let $0 = s_0 < s_1 < \cdots s_k$ the $\langle 1, s\rangle$-nondeficiency stages and note that $s_k < s$. t is a *$\langle 2, s\rangle$-nondeficiency stage* for B_2 if $t = 0$ or $t = s_i$ for some $0 < i \leq k$ and $\min\{B_2^{s_i} - B_2^{s_{i-1}}\} < \min\{B_2^s - B_2^{s_i}\}$.

By Lemma 1, if $s_0 < s_1 < \cdots$ are the 1-nondeficiency stages for B_1, then $B_2^{s_0} \subseteq B_2^{s_1} \subseteq \cdots$. Furthermore if s is fixed and $s_0 < s_1 < \cdots < s_k$ are the $\langle 1, s\rangle$-nondeficiency stages for B_1, then $B_2^{s_0} \subseteq B_2^{s_1} \subseteq \cdots \subseteq B_2^{s_k} \subseteq B_2^s$. We can now define the 2-nondeficiency approximation to B_2.

Definition 4. Let B_2 be c.e. relative to a c.e. set B_1 with the approximations as above. The *2-nondeficiency approximation* $\{\beta_2^s \mid s \in \mathbb{N}\}$ to B_2 is defined as follows. For $s = 0$, $\beta_2^0 = \emptyset$. For $s > 0$, let $t_2^s < s$ denote the largest $\langle 2, s\rangle$-nondeficiency stage. (By the comments above, $B_2^{t_2^s} \subseteq B_2^s$.) Let $\beta_2^s \preceq B_2^s$ with length $\min\{B_2^s - B_2^{t_2^s}\}$.

Since our real goal is to deal with 2-CEA sets, let B_1 be a c.e. set and let B_2 be c.e. relative to B_1 with the fixed approximations and terminology as above. Consider the 2-CEA set $B = B_1 \oplus B_2$. Abusing terminology slightly, we define the 2-nondeficiency approximation to B as $\{\beta^s \mid s \in \mathbb{N}\}$ by $\beta^0 = \emptyset$ and $\beta^s = \beta_1^s \oplus \beta_2^s$ for $s > 0$. (Since β_1^s and β_2^s need not have the same length, we introduce a formal symbol λ with the intention that $\beta_i^s(k) = \lambda$ indicates that β_i^s has no commitment concerning whether k is in or out of the approximated set. We extend the shorter string using λ symbols to make them have the same length. When comparing strings α and δ which may contain λ, we write $\alpha \preceq \delta$ if for all $k < |\alpha|$, if $\alpha(k) \neq \lambda$, then $k < |\delta|$ and $\alpha(k) = \delta(k)$.)

Definition 5. Let $B = B_1 \oplus B_2$ be as above. t is a *2-nondeficiency stage* for B if t is a 2-nondeficiency stage for B_2 (and hence is also a 1-nondeficiency stage for B_1). For a fixed s, $t < s$ is a *$\langle 2, s\rangle$-nondeficiency stage* for B if t is a $\langle 2, s\rangle$-nondeficiency stage for B_2 (and hence is also a $\langle 1, s\rangle$-nondeficiency stage for B_1). t is *2-correct* if $\beta^t \preceq B$. For any fixed s, $t < s$ is *$\langle 2, s\rangle$-correct* if $\beta^t \preceq \beta^s$.

From these definitions, it is easy to verify the analogs of Lemmas 1 and 2 for a 2-CEA set $B = B_1 \oplus B_2$ as above.

Lemma 4. *If $0 = s_0 < s_1 < \cdots$ are the 2-nondeficiency stages for $B_1 \oplus B_2$, then $\beta^{s_0} \preceq \beta^{s_1} \preceq \cdots$. Furthermore, for any fixed s, if $0 = s_0 < s_1 < \cdots s_k$ are the $\langle 2, s\rangle$-nondeficiency stages for $B_1 \oplus B_2$ then $s_k < s$ and $\beta^{s_0} \preceq \beta^{s_1} \preceq \cdots \preceq \beta^{s_k} \preceq \beta^s$.*

Lemma 5. *A stage t is a 2-nondeficiency stage for $B_1 \oplus B_2$ if and only if t is 2-correct. For $t < s$, t is a $\langle 2, s\rangle$-nondeficiency stage for $B_1 \oplus B_2$ if and only if t is $\langle 2, s\rangle$-correct.*

The analog of Lemma 3 fails in the case of 2-CEA sets $B_1 \oplus B_2$. It is easy to verify that if s is 2-correct, then there are infinitely many stages $t > s$ at which s is $\langle 2, t\rangle$-correct, but it is not true is general that if s is 2-correct, then s is $\langle 2, t\rangle$-correct at all stages $t > s$. However, we can get some additional correctness information which will be crucial to our proof.

Lemma 6. *If r is not 2-correct, then there are only finitely many stages $s > r$ such that r is $\langle 2, s\rangle$-correct.*

Proof. If r fails to be 2-correct because $\beta_1^r \not\preceq B_1$, then the conclusion follows immediately. Assume that $\beta_1^r \preceq B_1$ and therefore $B_2^r \subseteq B_2$. We must have $\beta_2^r \not\preceq B_2$ so there is a number $n < |\beta_2^r|$ such that $n \in B_2$ and $\beta_2^r(n) = 0$. Fix the least such n and fix a stage $t > r$ such that $\beta_1^t \preceq B_1$ and β_1^t contains the use for the enumeration of n into B_2. For all $s \geq t$, $\beta^r \not\preceq \beta^s$ and hence the conclusion follows. □

Lemma 7. *Let $r' < r < s$.*

1. *If r' is $\langle 2, r\rangle$-correct and r is $\langle 2, s\rangle$-correct, then r' is $\langle 2, s\rangle$-correct.*
2. *If r is $\langle 2, s\rangle$-correct and r' is not $\langle 2, s\rangle$-correct, then r' is not $\langle 2, r\rangle$-correct.*

Proof. To prove the first statement, notice that by assumption, $\beta^{r'} \preceq \beta^r$ and $\beta^r \preceq \beta^s$. Therefore, $\beta^{r'} \preceq \beta^s$ as required. The second statement follows immediately from the first statement. □

Lemma 8. *Let $r' < r < s$. If r' is $\langle 2, s\rangle$-correct but not $\langle 2, r\rangle$-correct, then r is not $\langle 2, t\rangle$-correct for any $t \geq s$ (and hence is not 2-correct).*

Proof. First, consider the situation at the B_1 level. Since r' is $\langle 2, s\rangle$-correct, and hence $\langle 1, s\rangle$-correct, $\beta_1^{r'} \preceq \beta_1^s$. Hence by Lemma 3, $\beta_1^{r'} \preceq \beta_1^r$. Therefore, in order to have $\beta^{r'} \not\preceq \beta^r$, we must have $\beta_2^{r'} \not\preceq \beta_2^r$.

Since $\beta_1^{r'} \preceq \beta_1^r$, it follows that $B_2^{r'} \subseteq B_2^r$. Therefore, to have $\beta_2^{r'} \not\preceq \beta_2^r$, we must have a number $n < |\beta_2^{r'}|$ enter B_2^r with oracle β_1^r. However, since $\beta_2^{r'} \preceq \beta_2^s$, we must have $n \notin B_2^s$. In order for n to leave the approximation to B_2 between stages r and s, a number $< |\beta_1^r|$ must enter B_1 between stages r and s. Therefore, $\beta_1^r \not\preceq \beta_1^s$ and hence by Lemma 3, β^r cannot be even $\langle 1, t\rangle$-correct for any $t \geq s$. □

This method of approximation using nondeficiency oracles can be continued by induction through the n-CEA hierarchy. Given the nondeficiency approximation $\{\beta^s \mid s \in \mathbb{N}\}$ for an n-CEA set $B = B_1 \oplus \cdots \oplus B_n$ and a set B_{n+1} which is c.e. in B (and of the form $W_e^B \oplus \mathbb{N}$ by convention), we fix an enumeration of B_{n+1} such that B_{n+1}^s consists of the elements enumerated with oracle β^s (and the convention than $2s+1$ enters B_{n+1} at stage $s+1$ with no oracle information). We define the $n+1$-nondeficiency stages by thinning the n-nondeficiency stages as above and we define the $\langle n+1, s\rangle$-nondeficiency stages by thinning the $\langle n, s\rangle$-nondeficiency stages as above. The $n+1$-nondeficiency approximation $\{\beta_{n+1}^s \mid s \in \mathbb{N}\}$ is defined (for $s > 0$) by letting t_{n+1}^s be the largest $\langle n+1, s\rangle$-nondeficiency stage and setting $\beta_{n+1}^s \preceq B_{n+1}^s$ with length $\min\{B_{n+1}^s - B_{n+1}^{t_{n+1}^s}\} + 1$. The analogs of Lemmas 4 and 5 hold by induction for all n, but the analogs of Lemmas 6 and 8 fail for $n \geq 3$.

We also note the following.

Remark 9. It is easy to show and well-known that finding an infinite homogeneous set for a computable stable coloring of pairs that is computable in a given degree is equivalent to finding an infinite subset of a given Δ_2^0 set or its complement computable in that degree.

Theorem 10. *There is a computable stable coloring of pairs that does not have a homogeneous set computable from a c-cappable 2-CEA set.*

Proof. By Remark 9, it suffices to show that there is a Δ_2^0 set D such that neither D nor its complement, $\overline{D}$, has a subset computable from a c-cappable 2-CEA degree. Let $\{\varphi_e : e \in \mathbb{N}\}$ be an effective enumeration of all computable partial functions. It suffices to satisfy the following requirements for all c.e. sets A, 2-CEA sets $B = B_1 \oplus B_2$ and computable partial functionals Γ (A and B come with fixed enumerations $\{A^s : s \in \mathbb{N}\}$ and $\{\beta_i^s : i \in \{1,2\} \ \& \ s \in \mathbb{N}\}$, respectively):

$$R_{A,B,\Gamma,D} : (|\Gamma(B)| = \infty \ \& \ \Gamma(B) \subset D \ \& \ A \text{ non-computable}) \Rightarrow \\ \exists C(C \leq_T A \ \& \ C \leq_T B \ \& \ \forall e(C \neq \varphi_e));$$

$$R_{A,B,\Gamma,\overline{D}} : (|\Gamma(B)| = \infty \;\&\; \Gamma(B) \subset \overline{D} \;\&\; A \text{ non-computable}) \Rightarrow$$
$$\exists C(C \leq_T A \;\&\; C \leq_T B \;\&\; \forall e(C \neq \varphi_e)).$$

Let $C_{A,B,\Gamma,D}$ and $C_{A,B,\Gamma,\overline{D}}$ be the sets constructed to satisfy these requirements. We subdivide each requirement into infinitely many subrequirements, one for each e:

$$R_{A,B,\Gamma,D,e} : (|\Gamma(B)| = \infty \;\&\; \Gamma(B) \subset D \;\&\; A \text{ non-computable}) \Rightarrow$$
$$(C_{A,B,\Gamma,D} \leq_T A \;\&\; C_{A,B,\Gamma,D} \leq_T B \;\&\; C_{A,B,\Gamma,D} \neq \varphi_e);$$

$$R_{A,B,\Gamma,\overline{D},e} : (|\Gamma(B)| = \infty \;\&\; \Gamma(B) \subset \overline{D} \;\&\; A \text{ non-computable}) \Rightarrow$$
$$(C_{A,B,\Gamma,\overline{D}} \leq_T A \;\&\; C_{A,B,\Gamma,\overline{D}} \leq_T B \;\&\; C_{A,B,\Gamma,\overline{D}} \neq \varphi_e).$$

We fix an effective ordering $\{R_i : i \in \mathbb{N}\}$ of all subrequirements, and refer to these as requirements. We say that R_i *has higher priority than* R_j if $i < j$. We also fix a computable partition $\{S_i : i \in \mathbb{N}\}$ of $\mathbb{N}$ in which each S_i is infinite.

Requirements will be in one of two states; a *waiting state* or a *diagonalization state.* At each stage s of the construction, we will have a subset P_i^s of $\{j : j \leq s\}$, and requirements may have *designated oracles* and *designated disruptors.* Action taken to satisfy requirements will be defined in terms of *configurations* and *permission.* Fix a requirement $R_i = R_{A,B,\Gamma,D,e}$. (Everything done henceforth for this requirement applies to requirements $R_{A,B,\Gamma,\overline{D},e}$, and is obtained by interchanging D and $\overline{D}$.) For fixed i, we say that t is *i-configured* at stage s if $t < s$ and $|\Gamma^t(\beta^t) \cap [i, \infty)| \geq i + 1$. We say that $x \in S_i$ is *A-permitted for i* at stage s if $\varphi_e^s(x) \downarrow= 0$, and $\min\{A^s - A^{s-1}\} \leq x$. Note that if t is i-configured at stage $r < s$ then t is i-configured at stage s.

The idea for satisfying R_i is as follows. At the first stage s at which some $x \in S_i$ is A-permitted for i, we place $x \in C^{s+1} = C^{s+1}_{A,B,\Gamma,D}$. If no such s exists, then either A is computable or $\{x \in S_i : \varphi_e(x) \downarrow= 0\}$ is finite; in the latter case, either φ_e is not total, or we will have an $x \in S_i$ such that $\varphi_e(x) \downarrow\neq 0 = C(x)$. So suppose that x exists. $C \leq_T A$ by a standard permitting argument. The argument that $C \leq_T B$ is a bit more complex, and assumes that $\Gamma(B)$ is an infinite subset of D. We first argue that there can be no $t < s$ such that β^t is correct and t is i-configured at stage s, else we would later place the designated disruptor for β^t into $\overline{D}$, thereby forcing $\Gamma(B) \not\subseteq D$. But if $\Gamma(B)$ is an infinite subset of D, then there must be a stage t at which β^t is correct and t is i-configured at stage t. By the previous sentence, we must have $t \geq s$. But t can be found computably from B, so any number in S_i that is placed into C through action for R_i must be in C^t. This provides a computation of C from a B oracle.

The Construction

Every parameter defined for a requirement at stage $s-1$ has the same value at stage s unless the parameter is redefined during stage $s-1$.

Stage 0*:* For all i, R_i is in the *waiting state*, $D^0 = \emptyset$, and $P_i^0 = \emptyset$.

Stage $s > 0$*:* We proceed by a subinduction on $\{i : i \leq s\}$. Fix i. We follow the first case below which applies.

Case 1: R_i is in the waiting state and some $x \in S_i$ is A-permitted at stage s. Fix the smallest such x, and place $x \in C^{s+1}_{A,B,\Gamma,D}$. If s is i-configured, set $P_i^{s+1} = P_i^s \cup \{s\}$. R_i is placed in the *diagonalization state* at all stages $> s$.

Case 2: R_i is in the diagonalization state and has some y as its *designated disruptor* at stage $s-1$, and there is a $j < i$ that has y as its designated disruptor at stage s. Fix t such that β^t is the designated oracle for R_i at stage s. Fix the smallest $z \in \Gamma^t(\beta^t) \cap [i, \infty)$ that is not a designated disruptor for any $k < i$ at stage s; z becomes the *designated disruptor* for R_i at stage s. (We will show later that such a z exists.) We specify that $z \in D^{s+1}$ if and only if $y \in D^s$. R_i remains in the *diagonalization state* at stage s with the same designated oracle. We now follow Case 3 if the conditions for following that case apply, and otherwise complete step i of the subinduction.

Case 3: R_i is in the diagonalization state at stage $s-1$ and there is an $r \in P_i^s$ such that β^r is $\langle 2, s\rangle$-correct. β^r becomes the *designated oracle* for R_i for the smallest such r, and we set $P_i^{s+1} = P_i^s \cap [0, r)$. Fix the smallest $z \in \Gamma^r(\beta^r) \cap [i, \infty)$ that is not a designated disruptor for any $k < i$ at stage s; z becomes the *designated disruptor* for R_i at stage s. (We will show later that such a z exists.) Place $z \in \overline{D}^{s+1}$ if R_i is a requirement of the form $R_{A,B,\Lambda,D,e}$, and place $z \in D^{s+1}$ if R_i is a requirement of the form $R_{A,B,\Lambda,\overline{D},e}$. R_i remains in the *diagonalization state* at stage s.

Case 4: Otherwise. If R_i is in the waiting state at the end of stage $s-1$ and s is i-configured, then we set $P_i^{s+1} = P_i^s \cup \{s\}$. In all cases, R_i retains the state it had at stage $s-1$, and nothing else is changed.

This completes the construction. Note that each requirement has at most one designated disruptor at a given stage, and that if t is i-configured, then $|\Gamma^t(\beta^t) \cap [i, \infty)| \geq i+1$. Hence the construction can always find a new designated disruptor when it needs to do so. Furthermore, only R_j for $j < i$ can cause designated disruptors for configurations for R_i to change, and any such R_j can do so at only finitely many stages. Thus each configuration for R_i has a final designated disruptor. These final designated disruptors will not be moved from D to $\overline{D}$ until some x is A-permitted for i, and if such a permission occurs at stage s, then such movement can only be caused by $t < s$, and each such t will cause a single designated disruptor to move, and will do so at most once. Hence D is Δ^0_2.

We note that Case 1 is followed for a given requirement R_i at most once, and that once it is followed at stage s, the sequence $\langle P_i^t : t \geq s \rangle$ is non-increasing, so has a limit. Thus Case 3 can be followed only finitely often for R_i. Case 2 is followed for R_i only when Case 3 is followed for R_j for some $j < i$. It follows by induction that Case 4 will be followed for R_i at all but finitely many stages.

We now show that R_i is satisfied for all i, and that this satisfaction is uniform for all subrequirements of the same master requirement R. Again, we only consider the case wherein $R_i = R_{A,B,\Gamma,D,e}$; a symmetric proof for the case in which $R_i = R_{A,B,\Gamma,\overline{D},e}$ is obtained by interchanging D and $\overline{D}$. Fix i. We may assume that $|\Gamma(B)| = \infty$, $\Gamma(B) \subseteq D$ and A is not computable, else the satisfaction of R_i is immediate. Under this assumption, we show that $C_{A,B,\Gamma,D} \leq_T A, B$ and $C_{A,B,\Gamma,D} \neq \varphi_m$ for all m.

We begin by showing that $C = C_{A,B,\Gamma,D} \neq \varphi_e$; we may assume that φ_e is total, else there is nothing to show. We also fix a stage s such that all R_j for $j < i$ follow Case 4 of the construction at all stages $t \geq s$. It follows that each configuration β^r for such an R_j, has a designated disruptor at stage $t \geq \max\{r, s\}$ if and only if it has one at stage $\max\{r, s\}$, and that any designated disruptor at stage $\max\{r, s\}$ remains the designated disruptor at all later stages. As $|\Gamma(B)| = \infty$, it follows from Lemma 5 that there are infinitely many 2-correct stages t, and at all but finitely many of these stages, $|\Gamma^t(\beta^t)| \cap [i, \infty) \geq i+1$.

First assume that R_i is in the waiting state at all stages. As $S_i \cap S_j = \emptyset$ if $i \neq j$, we see that no number in S_i is placed in C at any stage. Thus if there is an $x \in S_i$ such that $\varphi_e(x) \neq 0$, then we will have an x such that $\varphi_e(x) \neq 0 = C(x)$, so $\varphi_e \neq C$. Otherwise, there are arbitrarily large $x \in S_i$ such that $\varphi_e(x) = 0$, so as A is not computable, there would be a stage t at which some x is A-permitted for i, so we would place R_i in the diagonalization stage at stage t, contrary to our assumption. Thus this case cannot occur.

Now assume that R_i is in the diagonalization state at some stage t. Then some x such that $\varphi^s(x) = 0$ is placed in C^{s+1}, so again we see that $\varphi_e \neq C$.

A standard permitting argument shows that for all x, for the least stage s such that $A^s \upharpoonright x = A \upharpoonright x$, $x \in C^{s+1}$ if and only if $x \in C$. Thus $C \leq_T A$.

We now describe how to compute C using a B oracle. Uniformly in x, we can effectively find the unique j such that $x \in S_j$. Fix x and the unique j such that $x \in S_j$. If $R_i \neq R_{A,B,\Gamma,D,m}$ for any m, then $x \notin C$. Hence, without loss of generality, we may suppose that $x \in S_i$, i.e., that $j = i$. Use a B oracle to find the least 2-correct stage t such that $|\Gamma^t(\beta^t)| \geq i + 1$. By our assumptions and the fact that there are infinitely many 2-correct stages, such a stage must exist.

We claim that $x \in C$ if and only if $x \in C_{t+1}$. For suppose that $x \in C$. Then x is placed into C at some stage s at which x is A-permitted for i. If $s \leq t$, then $x \in C_{t+1}$ as required. Suppose $t < s$ and we derive a contradiction. Since R_i remains in the waiting state until x is placed into C and since t is i-configured, $t \in P_i^s$. We show that eventually β^t is chosen as the designated oracle for R_i. To see why, let $s' \geq s$ be the first stage at which R_i picks a designated oracle and let β^r (for some $r \in P_i^{s'} = P_i^s$) be the designated oracle chosen. (Since t is $\langle 2, s' \rangle$-correct for infinitely many stages s' and since $t \in P_i^s$, there must be such a stage.)

We claim that $t \leq r$. For a contradiction, suppose that $r < t$. Then $r < t < s'$ and r is $\langle 2, s' \rangle$-correct. By Lemma 8, if r is not $\langle 2, t \rangle$ correct, then t is not 2-correct. Therefore, r must be $\langle 2, t \rangle$-correct and hence r is 2-correct since $\beta^r \preceq \beta^t \preceq B_1 \oplus B_2$. This contradicts the minimality property of t.

If $r = t$, then we have shown that β^t is eventually chosen as the designated oracle for R_i. If $t < r$, then r cannot be 2-correct. (If r is 2-correct, then $\beta^t \preceq \beta^r \preceq \beta^{s'}$. However, if t is $\langle 2, s' \rangle$-correct, then we would have chosen β^t as the designated oracle at stage s'.) Therefore, there must be a later stage s'' and a index $r' < r$ such that $r' \in P_i^{s''} = P_i^s \cap [0, r)$ such that r' is $\langle 2, s'' \rangle$-correct. As above, $t \leq r'$. Since r' was not chosen as stage s', r' is not $\langle 2, s' \rangle$-correct and hence by Lemma 7 (since $r' < r < s'$), r' is also not $\langle 2, r \rangle$-correct. We can now apply Lemma 8 to the stages $r' < r < s''$ to conclude that r is not $\langle 2, p \rangle$-correct for any $p \geq s''$. (Therefore, it was safe to define $P_i^{s''} = P_i^s \cap [0, r')$.) If $t = r'$, then we have shown that β^t is eventually chosen as the designated oracle. If $t < r''$, then we repeat this argument defining successive new designated oracles until β^t is chosen.

Once β^t is chosen as the designated oracle, it remains the designated oracle forever (by Lemma 8 and the arguments given in the previous paragraph). We have shown that β^t has a final designated disruptor y for i. By construction, $y \in \overline{D}$ but (since β^t is 2-correct) $\Gamma(B;y) = \Gamma^t(\beta^t) = 1$. Therefore, $\Gamma(B) \nsubseteq D$ contrary to our assumption. We conclude that $x \in C$ if and only if $X \in C^{t+1}$, so $C \leq_T B$. □

The proof of Theorem 10 does not generalize to the 3-CEA case. The problem for the 3-CEA case lies in the fact that we may have several pairwise incompatible approximations to B, each of which looks correct at infinitely many stages; this prevents us from making D a Δ^0_2 set. We have not yet looked at the non-stable case. Thus we conclude with some questions.

Question 1. For which integers n, if any, is it true that every computable stable 2-coloring of ω^2 has a c-cappable n-CEA homogeneous set?

We have a similar question for the non-stable case.

Question 2. For which integers n, if any, is it true that every computable 2-coloring of ω^2 has a c-cappable n-CEA homogeneous set?

References

[1] Ambos-Spies, K., Jockusch, C. G. Jr., and Shore, R. A., and Soare, R. I., *An algebraic decomposition of the recursively enumerable degrees and the coincidence of several degree classes with the promptly simple degrees*, Trans. Amer. Math. Soc. **281** (1984), 109-128.

[2] Ambos-Spies, K., Lempp, S., and Lerman, M., *Lattice embeddings into the r.e. degrees preserving* 0 *and* 1, J. London Math. Soc. (2) **49** (1994), 1-15.

[3] Cholak, Peter A., Jockusch, Carl G., and Slaman, Theodore A., *On the strength of Ramsey's Theorem for pairs*, J. Symbolic Logic **66** (2001), 1-55.

[4] Ding, D. and Qian L., *Lattice embedding into d-r.e. degrees preserving 0 and 1*, In: Proceedings of the Sixth Asian Logic Conference (Chong, Feng, Ding, Huang and Yasugi eds.), World Scientific Press, Singapore, 1998, pp. 67-81.

[5] Downey, R., *D-r.e. degrees and the non-diamond theorem*, Bull. London Math. Soc. **21** (1989), 43-50.

[6] Downey, R., Hirschfeldt D., Lempp, S. and Solomon, D. R., *A* Δ^0_2 *set with no infinite low set contained in it or its complement*, J. Symbolic Logic **66**(3) (2001), 1371-1381.

[7] Hirschfeldt, Denis R., Jockusch, Jr., Carl G., Kjos-Hanssen, Bjørn; Lempp, Steffen, and Slaman, Theodore A., *The strength of some combinatorial principles related to Ramsey's Theorem for pairs*, In: Computational Prospects of Infinity, Part II: Presented Talks, World Scientific Press, Singapore, 2008, pp. 143-161.

[8] Jockusch, Jr., Carl G., and Shore, Richard A., *Pseudo-Jump Operators I: the r.e. case*, Trans. Amer. Math. Soc. **275** (1983), 599-610.
[9] Soare, Robert I., Recursively Enumerable Sets and Degrees, Perspectives In Mathematical Logic, Springer-Verlag, Berlin-Heidelberg-New York, 1987.

COMPUTABLE DOWD-TYPE GENERIC ORACLES

Masahiro Kumabe
Faculty of Liberal Arts, The Open University of Japan,
Wakaba 2-11, Mihama-ku, Chiba-city 261-8586, Japan
kumabe@ouj.ac.jp

Toshio Suzuki*
Department of Mathematics and Information Sciences,
Tokyo Metropolitan University,
Minami-Ohsawa, Hachioji, Tokyo 192-0397, Japan
toshio-suzuki@tmu.ac.jp

In [1] (*Proc. ALC 10*), we study the following property $(*)$ of an oracle X. $(*)$ "For every positive integer r, X is r-generic in the sense of Dowd (1992, *Inf. Comput.*)" (that is, for each r, the forcing complexity of the r-query tautologies with respect to X is bounded by a polynomial; this property is different from r-genericity of arithmetical forcing). In [1], we show that if X is Schnorr random then X has the property $(*)$, and that the property $(*)$ is conserved through a certain type of p-time bounded-truth-table reduction. In the current article, we show (1) and (2) below. (1) There exists a primitive recursive oracle X that has the property $(*)$; (2) Every Turing degree contains an oracle X that has the property $(*)$. The results (1) and (2) are extensions of results in [2] (Suzuki, 2002, *Inf. Comput.*) about 1-genericity in the sense of Dowd. (2) is an affirmative solution to Problem 4 in [2].

Mathematics Subject Classification: 68Q15, 03D15.

Keywords: Dowd-type generic oracle, computability, effective randomness.

1. Introduction

The concept of Dowd-type generic oracle is a sort of resource-bounded version of Cohen-type generic oracle. In the definition of Dowd-type genericity, we bound the sizes of forcing conditions associated to a given oracle, and, on the other hand, we do not bound time-complexity of open dense sets

*Corresponding author. This work was partially supported by Japan Society for the Promotion of Science (JSPS) KAKENHI (B) 19340019 and (C) 22540146.

associated to an oracle. We consider the relativized propositional calculus (RPC), a system of Boolean formulas with query symbols to a given oracle. If F is a formula of RPC such that the occurrence of query symbols in F is just r, where r is a positive integer, then F is called an r-query formula. If, in addition, F is a tautology with respect to a given oracle X then F is called an r-query tautology with respect to X. For a positive integer r, an oracle D is called an r-Dowd oracle if every r-query tautology with respect to D is forced by a finite portion S of D such that the size of S (the cardinality of its domain) is polynomial in the length of F. In our former paper [1], we show that every Schnorr random oracle is r-Dowd for every positive integer r, and we show that the property of being r-Dowd for every positive integer r is conserved through a certain type of p-time bounded-truth-table reduction.

The aim of this article is to prove the following theorem.

Theorem 5.1 (Main theorem) (1) *There exists a primitive recursive oracle D such that for every positive integer r, D is r-Dowd.*

(2) *Every Turing degree contains an oracle D such that for every positive integer r, D is r-Dowd.*

The main theorem extends our former result below.

Proposition 1.1. *[2, Theorem 2, 3] (1) There exists a primitive recursive oracle D such that D is 1-Dowd.*

(2) Every Turing degree contains an oracle D such that D is 1-Dowd.

The uniqueness of the minimum forcing condition plays an important role in the proof of Proposition 1.1. More precisely, the following holds.

Proposition 1.2. *(Dowd, [3, Lemma 9]) If a 1-query formula F is a tautology with respect to some oracle X, then there exists a forcing condition S_F such that for every oracle A, the necessary and sufficient condition for "F is a tautology with respect to A" is "A is an extension of S_F". In Dowd's terminology, F* specifies *S_F.*

By a counting argument using the uniqueness of the specified forcing condition, we get the following lower bound of probability. Suppose that p is a fixed polynomial, n is a positive integer and that S is a forcing condition whose domain is $\{0,1\}^n$; to be more precise, the domain is a certain coding of $\{0,1\}^n$. In addition, assume that S is 1-Dowd with respect to p; In this section, we denote the above assumption by the phrase "S is a survivor at stage n". Now, we randomly choose a forcing condition S' such that the domain of S' is $\{0,1\}^{n+1}$, and S' extends S, where we identify each $u \in \{0,1\}^n$ with $0u \in \{0,1\}^{n+1}$. Then, under certain assumptions, the probability of S' being 1-Dowd with respect to p is at least $1 - 1/2^{n+2}$. (See [2,3].)

Therefore we get an infinite binary tree T such that every node S of height n is a survivor at stage n, and the two child nodes of S are the first and the second elements of the following set with respect to length-lexicographic order: $\{S' : S'$ is a survivor at stage $n+1$ and S' is an extension of $S\}$. Hence, every Turing degree is coded by an infinite branch of this tree. In particular, the infinite branch of T corresponding to the Turing degree 0 is a primitive recursive 1-Dowd oracle. For detail of the proof of Proposition 1.1, consult [2].

In summary, in the case of 1-query, with a counting argument based on Proposition 1.2, we know that if a forcing condition S is a survivor at stage n then S has at least two (in fact, many) "child nodes" that are survivors at stage $n+1$. With this property, we get a binary tree whose nodes are survivors, and the "least" infinite branch of this tree is a primitive recursive 1-Dowd oracle.

However, in the case where $r \geq 2$, a counter part of Proposition 1.2 does not hold. An informal counter example is as follows: let n be a positive integer and consider a 2-query formula F asserting that for a given oracle X, the value of X at the string 000 is not the same as that at 001. Then there exist two minimal forcing conditions that forces F; one has value $0, 1$ at $000, 001$ respectively, and the other has value $1, 0$ at $000, 001$ respectively.

Since we do not have a counter part of Proposition 1.2, the above argument on "survivors" does not work. We have only a weaker version of Proposition 1.2. Suppose $r \geq 2$. For each oracle A, we define a certain subset of r-query tautologies with respect to A. We call them *nice r-query tautologies with respect to A*. Then we get a revised version of Proposition 1.2: If F is a nice r-query tautology with respect to some oracle X and if T is a certain special type of finite portion of X, then there is a unique minimal forcing condition S_T such that the union of S_T and T forces F. (See [4,5].) In section 3, we review the concept of nice r-query tautology.

In this article, we investigate a property "survivor in the strong sense". We show that if a forcing condition S is a survivor in the strong sense at stage n then S has an extension that is a survivor in the strong sense at stage $n+1$. With this property, we show that for every integer $r \geq 2$, there exists a primitive recursive r-Dowd oracle (section 4). With a similar argument, we show the main theorem (section 5).

Informally speaking, the concept of "S is a survivor in the strong sense" is defined as follows. We fix a polynomial p and a number-theoretical function f such that $f(n)$ is sufficiently larger than n for every positive integer n. Suppose that S is a forcing condition whose domain is (a certain coding of) $\{0,1\}^n$. The statement "S is a survivor in the strong sense" denotes that (1) and (2) below hold.

(1) S is r-Dowd with respect to p.

(2) We randomly choose a forcing condition T of the property (2a). Then the conditional probability of T having property (2b) is high enough.

2a T is a forcing condition whose domain is $\{0,1\}^{f(n)}$, and T extends S.
2b For every i such that $n+1 \leq i \leq f(n)$, the restriction of T (its domain) to $\{0,1\}^i$ is r-Dowd with respect to p.

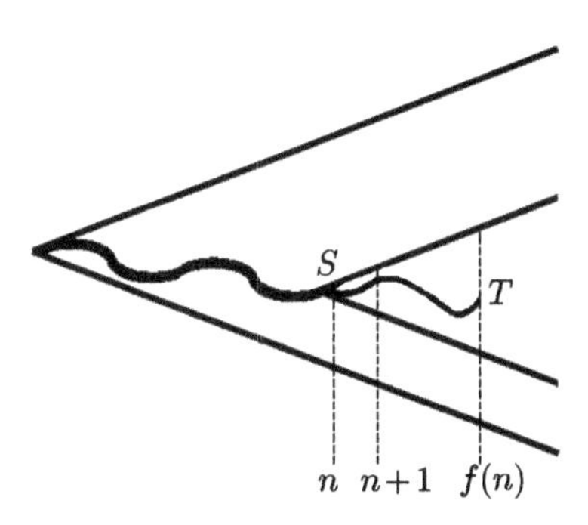

Fig. 1. S survives in the strong sense at stage n

The main theorem is an affirmative solution to Problem 4 in [2].

We close this introduction section with an open question. In [1], we show that if an oracle X is Schnorr random then for every positive integer r, X is r-Dowd. Question 1 below asks whether the same thing as above hold for Lutz's resource-bounded randomness in place of Schnorr randomness.

Question 1. *Is there a primitive recursive function $t(n)$ of the following property? "For every oracle X, if X is $t(n)$-random in the sense of Lutz then for every positive integer r, X is r-Dowd."*

It is known that there exists a primitive recursive $t(n)$-random set in the sense of Lutz. [12, p.199] Therefore, an affirmative answer for Question 1 would be an alternative proof of the main theorem. Question 1 is suggested by K. Ambos-Spies in personal communication. For Lutz's resource-bounded randomness, see [7,8].

2. Notation

2.1. *Strings and sets*

We let $\{0,1\}^*$ denote the set of all finite binary strings. For each natural number n, we let $\{0,1\}^n$ denote the set of all binary strings of length n. We represent the empty string by λ. For each natural number n, we let $z(n)$ denote the $(n+1)$st string in length-lexicographic order. For example, $z(0) = \lambda, z(1) = 0, z(2) = 1, z(3) = 00, z(4) = 01$. For each string u, we represent its length by $|u|$. For a set A, we let $|A|$ denote the cardinality of A. We believe that this abuse of the symbol "$|\quad|$" does not confuse the reader. A subset of $\{0,1\}^*$ is called an *oracle*. We identify an oracle with its characteristic function.

For a function f, we denote its domain by dom$f$. For a subset A of dom$f$, we let $f \upharpoonright A$ denote the restriction of f to A, that is, its domain is A and for every $x \in A$, $(f \upharpoonright A)(x) = f(x)$. When we apply this notation to the case where f is an oracle X, we regard X as its characteristic function.

Suppose that f is a function, n is a natural number and $\{z(k) : k < n\}$ is a subset of dom$f$. Then we denote $f \upharpoonright \{z(k) : k < n\}$ by $f \upharpoonright z(n)$.

2.2. *Probability*

The set of all oracles is called the *Cantor space*. We regard the Cantor space as a probability space with the standard Lebesgue measure. In other words, for an oracle X and a natural number n, it holds that Prob$[X(n) = 0] =$ Prob$[X(n) = 1] = 1/2$, where Prob denotes probability. For events E_1 and E_2, we let Prob$[E_1|E_2]$ denote the conditional probability of E_1 assuming that E_2 occurs.

2.3. *Dowd-type generic oracles*

This subsection is a summary of the subsection of the same title in [1]. A function is called *a forcing condition* if its domain is a finite subset of $\{0,1\}^*$ and its range is a subset of $\{0,1\}$. By adding a set of query symbols $\{\xi^1, \xi^2, \xi^3, \cdots\}$ to the usual propositional calculus, we get *the relativized propositional calculus* [3]. For each n, the symbol ξ^n is an n-ary connective. We interpret ξ^n as follows. For an oracle A and each positive integer n, we define an n-ary function A^n so that, for each $j < 2^n$,

$$A^n(\text{ the } (j+1)\text{st } n\text{-bit string }) = A(\text{ the } (j+1)\text{st string }).$$

Thus, for example,

$$A^3(000) = A(\lambda),\ \ A^3(001) = A(0),\ \ A^3(010) = A(1),\ \ A^3(011) = A(00),$$
$$A^3(100) = A(01),\ A^3(101) = A(10),\ A^3(110) = A(11),\ A^3(111) = A(000).$$

Then, the following relation holds.

$$A^{n+1}(0, x_1, \cdots, x_n) = A^n(x_1, \cdots, x_n)$$

For a given oracle A, we interpret ξ^n as A^n.

If S is a forcing condition and F is a tautology with respect to every oracle X that is an extension of S, then we say that S *forces* F. Suppose that r is a positive integer. A formula of the relativized propositional calculus is called *an r-query formula* if it is of the following form.

$$[(q_0^{(1)} \Leftrightarrow \xi^n(q_1^{(1)}, \cdots, q_n^{(1)}) \wedge \cdots \wedge (q_0^{(r)} \Leftrightarrow \xi^n(q_1^{(r)}, \cdots, q_n^{(r)}))] \Rightarrow H,$$

where each of $q_j^{(i)}$ is a propositional variable and H is a query free formula. H may have occurrences of each of $q_j^{(i)}$, and may have occurrences of other propositional variables.

For a relativized formula F, we let $\ell(F)$ denote the length of F, that is, the total number of occurrences of propositional variables, constants (0 and 1), logical connectives, query symbols (ξ^n), punctuation marks and parentheses.

An oracle D is called *an r-Dowd oracle* if there exists a polynomial p of the following property: For every r-query formula F that is a tautology with respect to D, there exists a forcing condition S such that S is a sub-function of D, the cardinality of the domain of S is at most $p(\ell(F))$ and such that S forces F.

For more formal treatment of Dowd-type genericity, see [1–3,6].

3. Review of the Former Results

In sub-section 3.1, we cite some definitions and results from [1, section 3]. In sub-section 3.2, we present remarks on the former results.

3.1. *Review of our former paper*

We begin with defining "nice r-query formula".

Definition 3.1. [4,5]

(1) Suppose that n is a positive integer and that each of $\alpha_1, \cdots, \alpha_n$ is either a propositional variable or a constant (0 or 1). Then, the formula

$$\xi^n(\alpha_1, \cdots, \alpha_n)$$

is called a *query*.

(2) Two queries $\xi^n(\alpha_1, \cdots, \alpha_n)$ and $\xi^n(\beta_1, \cdots, \beta_n)$ are *disjoint* if there exists $i \in \{1, \cdots, n\}$ such that both α_i and β_i are constants and that $\alpha_i \neq \beta_i$. Note that if they are disjoint then the following sets of strings are disjoint.

$$\{x_1 \cdots x_n \in \{0,1\}^n : \forall i \, (\alpha_i \text{ is a constant} \to x_i = \alpha_i)\}, \text{ and}$$
$$\{x_1 \cdots x_n \in \{0,1\}^n : \forall i \, (\beta_i \text{ is a constant} \to x_i = \beta_i)\}.$$

(3) Suppose that $r \geq 2$. A relativized formula F is called *a nice r-query formula* if it is of the form

$$\left[(q_0^{(1)} \Leftrightarrow \xi^n(\cdots)) \wedge \cdots \wedge (q_0^{(r)} \Leftrightarrow \xi^n(\cdots))\right] \Rightarrow H,$$

where each of $q_0^{(1)}, \cdots, q_0^{(r)}$ is a propositional variable, H is a query-free formula, each $\xi^n(\cdots)$ is a query, and the queries are pairwise disjoint.

Example 3.1. Let F_1 be the following formula.

$$\left[(q_0^{(1)} \Leftrightarrow \xi^3(0,0,q_3^{(1)})) \wedge (q_0^{(2)} \Leftrightarrow \xi^3(0,1,q_3^{(2)})) \wedge (q_0^{(3)} \Leftrightarrow \xi^3(1,0,q_3^{(3)}))\right]$$
$$\Rightarrow \left[(q_0^{(1)} \vee q_0^{(2)}) \wedge q_0^{(3)}\right].$$

In addition, let F_2 be the following formula.

$$\left[(q_0^{(1)} \Leftrightarrow \xi^3(0,0,q_3^{(1)})) \wedge (q_0^{(2)} \Leftrightarrow \xi^3(0,1,q_3^{(2)})) \wedge (q_0^{(3)} \Leftrightarrow \xi^3(q_3^{(1)},1,q_3^{(3)}))\right]$$
$$\Rightarrow \left[(q_0^{(1)} \vee q_0^{(2)}) \wedge q_0^{(3)}\right].$$

Then, F_1 is a nice 3-query formula. On the other hand, F_2 is not a nice 3-query formula because the second query and the third query are not disjoint.

Next, we define the concept of "to force partially".

Definition 3.2. Suppose that n is a positive integer and that X is a forcing condition.

(1) In sub-section 2.3, for each oracle A, we define the n-ary function A^n. In the same way, we define an n-ary Boolean (partial) function X^n. That is, For each $j < 2^n$ such that $z(j) \in \text{dom}X$,

$$X^n(\text{ the } (j+1)\text{st } n\text{-bit string }) = X(\text{ the } (j+1)\text{st string }).$$

(2) For each string u of length n, letting $u = u_1 \cdots u_n$, we define meta-symbols $*_{u,1}, \cdots, *_{u,n}$ as follows. For each $j = 1, \cdots, n$,

$$*_{u,j} \text{ denotes } \begin{cases} \text{ (empty expression), if } u_j = 1, \\ \neg \text{ (negation symbol), otherwise.} \end{cases}$$

In addition, if $u \in \mathrm{dom} X^n$, we define a meta-symbol and $*_u$ as follows.

$$*_u \text{ denotes } \begin{cases} \text{ (empty expression), if } X^n(u) = 1, \\ \neg \text{ (negation symbol), otherwise.} \end{cases}$$

(3) Suppose that q is a propositional variable, $\xi^n(\alpha_1, \cdots, \alpha_n)$ is a query in the sense of Definition 3.1 (1). Let A be the set of all strings u such that $|u| = n$ and such that the formula $*_{u,1}\ \alpha_1 \wedge \cdots \wedge *_{u,n}\ \alpha_n$ is satisfiable. Assume that $A \subseteq \mathrm{dom} X^n$. Then, we let meta-symbol

$$[\![q \Leftrightarrow \xi^n(\alpha_1, \cdots, \alpha_n)]\!]^X$$

denote the conjunction of all the formulas of the form

$$(*_{u,1}\ \alpha_1 \wedge \cdots \wedge *_{u,n}\ \alpha_n) \Rightarrow *_u\ q,$$

where $u \in A$.

(4) Suppose that $r \geq 2$ is a natural number and that F is a nice r-query formula of the form

$$\big[(q_0^{(1)} \Leftrightarrow \xi^n(\alpha_1^{(1)}, \cdots, \alpha_n^{(1)})) \wedge (q_0^{(2)} \Leftrightarrow \xi^n(\alpha_1^{(2)}, \cdots, \alpha_n^{(2)})) \wedge \cdots$$
$$\wedge\, (q_0^{(r)} \Leftrightarrow \xi^n(\alpha_1^{(r)}, \cdots, \alpha_n^{(r)}))\big] \Rightarrow H.$$

In addition, suppose that S is a forcing condition and that the domain of X^n contains all strings relevant to the queries of F from the second query onward. Then, the sentence "*S partially forces F with respect to X*" denotes that S forces the following 1-query formula.

$$\big[(q_0^{(1)} \Leftrightarrow \xi^n(\alpha_1^{(1)}, \cdots, \alpha_n^{(1)})) \wedge [\![q_0^{(2)} \Leftrightarrow \xi^n(\alpha_1^{(2)}, \cdots, \alpha_n^{(2)})]\!]^X \wedge \cdots$$
$$\wedge\, [\![q_0^{(r)} \Leftrightarrow \xi^n(\alpha_1^{(r)}, \cdots, \alpha_n^{(r)})]\!]^X\big] \Rightarrow H.$$

Note that each of the following predicates $\Psi_1(F)$ and $\Psi_2(S, F, X)$ is primitive recursive.

$\Psi_1(F) \equiv$ "F is a nice r-query formula",

$\Psi_2(S, F, X) \equiv$ "S partially forces F with respect to X".

Remark: For more formal treatments of Definition 3.1 and Definition 3.2, see [4, §4.2], in which "a nice r-query formula" is called "a disentangled r-query formula", and the technical term "partially forces" is not explicitly used. In [2], a symbol $r\mathrm{DO}_1$ is used instead of $r\mathrm{GEN}_1$. The motive for these names are "r-**D**owd **o**racle" and "r-**gen**eric oracle in the sense of Dowd", respectively.

Definition 3.3. [4, p.38–39] Suppose that $r \geq 2$.

(1) Suppose that F is a relativized formula having at least one occurrence of a query symbol. Then, the dimension of F denotes the maximum natural number n such that the query symbol ξ^n has at least one occurrence in F. The dimension is denoted by $\mathrm{dim}F$.
(2) For each positive integer n, we let $\mathrm{Func}(n)$ denote the set of all functions from $\{z(0), z(1), \cdots, z(2^n - 1)\}$ to $\{0, 1\}$.
(3) For each positive integers r, k, c, n, we let $r\mathrm{GEN}_1(k, c, n)$ denote the set of all functions $S \in \mathrm{Func}(n)$ with the following property: For every r-query formula F such that

$$\begin{cases} \ell(F) \geq c, \quad \mathrm{dim}F = n, \quad F \text{ is nice, and} \\ F \text{ is a tautology with respect to S,} \end{cases}$$

there exists a function S' such that

$$\begin{cases} S' \text{ is a sub-function of } S, \quad \left|\mathrm{dom}S'\right| \leq \ell(F)^k, \text{ and} \\ S' \text{ partially forces } F \text{ with respect to } S, \end{cases}$$

where $\ell(F)$ is the length of F (see sub-section 2.3).
(4) For each positive integers r, k, c, we let $r\mathrm{GEN}_1(k, c)$ denote the set of all oracles X such that

$$\forall n \geq 1, X \restriction z(2^n) \in r\mathrm{GEN}_1(k, c, n),$$

where the symbol $\restriction$ denotes restriction of a function to a given subset of its domain (see sub-section 2.1).
(5) We define $r\mathrm{GEN}_1$ as follows, where the union extends over all positive integers k and c.

$$r\mathrm{GEN}_1 = \bigcup_{k,c \geq 1} r\mathrm{GEN}_1(k, c).$$

Lemma 3.1. *[4, p.41] Suppose that r, k, c and n are positive integers. Suppose that $r \geq 2$, and that k, c are large enough. Let N denote 2^n. Then, the following holds.*

$$\left|\mathrm{Func}(n) - r\mathrm{GEN}_1(k, c, n)\right| \leq 2^{N-(n+2)} \tag{1}$$

Lemma 3.2. *[4, Theorem 4.5] For each $r \geq 2$, the set of all r-Dowd oracles is equal to $r\mathrm{GEN}_1$.*

3.2. *Comments on our former paper*

Remark on Definition 3.1: The concept of a "nice r-query formula" in [1, p.331] is not exactly equivalent to that of a "disentangled r-query formula" in [4]. The former is a necessary condition for the latter. According to this difference, the class $r\mathrm{GEN}_1(k,c)$ defined in [1] is not equal to $r\mathrm{GEN}_1(k,c)$ defined in [4]. The former is a subclass of the latter. However, all the proofs of Proposition 4.1–4.4 in [4, pp.39–44] work under the setting of [1], too. In the current article, we adopt the concept of a "nice r-query formula" in [1].

Remark on Definition 3.3: The definition of $\mathrm{Func}(n)$ in [1] has a typo. That in the current article is right.

Remark on Lemma 3.1: We can replace the phrase "k, c are large enough" by more explicit one. Let c_1 be the total number of symbols of the relativized propositional calculus that are neither variables nor query symbols. That is, c_1 is the number of constants (0 and 1), logical symbols ($\Rightarrow$, etc.) and parentheses. The exact value of c_1 depends on our syntactic convention on the relativized propositional calculus. Then, Lemma 3.1 holds for the phrase "$k \geq 3$ and $c \geq c_1$" in place of "k, c are large enough". This is shown by carefully reading [4, p.41]. However, in this article, we do not explicitly use this result.

The other remark on the former results is the following.

Proposition 3.1. *Suppose that r, k, c, n are natural numbers such that $r \geq 2$, $k \geq 1$ and $c \geq 2^n$. And, suppose that S is a forcing condition whose domain is $\{z(0), z(1), \cdots, z(2^n - 1)\}$. Then, $S \in r\mathrm{GEN}_1(k,c,n)$.*

Proof. Suppose that F is a nice r-query formula of dimension n such that $\ell(F) \geq c$ and that F is a tautology with respect to F. Then, S forces F, and hence S has a sub-function S' that partially forces F with respect to S. And, the cardinality of the domain of S' is less than $2^n \leq c \leq c^k \leq \ell(F)^k$. Hence, S belongs to $r\mathrm{GEN}_1(k,c,n)$. □

4. The Case where r is Fixed

In this section, we investigate the case where we fix the number r of query symbols in a formula. We hope that this section assists the reader in understanding the next section. The aim of this section is to show the following lemma.

Lemma 4.1. *Let $r \geq 2$ be a natural number. Then, there exists a primitive recursive oracle D such that D is r-Dowd.*

Throughout this section, for each natural number n, let

$$f(n) = 2^n + n + 1.$$

The following is a "one-step lemma" in our proof of Lemma 4.1.

Lemma 4.2. *Fix positive integers c and k such that for every positive integers $r \geq 2$ and n, the equation (1) of Lemma 3.1 holds.*

Let $r \geq 2$ be a natural number. For a forcing condition S and a positive integer n, we let $R(S, n)$ denote the following requirement.

$$R(S,n): \text{“}S \in r\mathrm{GEN}_1(k,c,n) \text{ and } \mathrm{Prob}[E_1^n | E_2^{S,n}] \geq \frac{1}{2}(1 + \frac{1}{2^{n+1}})\text{”},$$

where the events E_1^n and $E_2^{S,n}$ are defined as follows. We randomly choose a forcing condition $T : \{z(0), z(1), \cdots, z(2^{f(n)} - 1)\} \to \{0, 1\}$. Then,

$$E_1^n : \text{“}n+1 \leq \forall i \leq f(n) \;\; T \restriction z(2^i) \in r\mathrm{GEN}_1(k,c,i)\text{”},$$

$$E_2^{S,n} : \text{“ } T \text{ extends } S\text{”}.$$

Suppose that n is a positive integer, S is a forcing condition such that its domain is $\{z(0), z(1), \cdots, z(2^n - 1)\}$ and suppose that $R(S, n)$ holds.

Then, there exists a forcing condition S' such that S' extends S, the domain of S' is $\{z(0), z(1), \cdots, z(2^{n+1} - 1)\}$ and that $R(S', n+1)$ is satisfied.

Proof. Let E_3^{n+1} be the following event with respect to a randomly chosen forcing condition $T : \{z(0), z(1), \cdots, z(2^{f(n+1)} - 1)\} \to \{0, 1\}$.

$$E_3^{n+1} : \text{“}f(n) + 1 \leq \forall i \leq f(n+1) \;\; T \restriction z(2^i) \in r\mathrm{GEN}_1(k,c,i)\text{”}.$$

Note that $E_2^{S,n+1}$ denotes the event of "T extends S". Then, the following holds, where we identify an event with the set of all forcing conditions T for which the event occurs.

$$\begin{aligned}
&\mathrm{Prob}[E_2^{S,n+1}] \times \mathrm{Prob}[E_3^{n+1} | E_2^{S,n+1}] \\
=&\mathrm{Prob}[E_3^{n+1} \cap E_2^{S,n+1}] \\
\geq&\mathrm{Prob}[E_2^{S,n+1} - \bigcup_{i=f(n)+1}^{f(n+1)} \{T : T \restriction z(2^i) \notin r\mathrm{GEN}_1(k,c,i)\}] \\
\geq&\mathrm{Prob}[E_2^{S,n+1}] - \sum_{i=f(n)+1}^{f(n+1)} \mathrm{Prob}[T \restriction z(2^i) \notin r\mathrm{GEN}_1(k,c,i)].
\end{aligned}$$

Therefore, by Lemma 3.1, the following holds.

$$\mathrm{Prob}[E_2^{S,n+1}] \times \mathrm{Prob}[E_3^{n+1}|E_2^{S,n+1}] \geq \mathrm{Prob}[E_2^{S,n+1}] - \sum_{i=f(n)+1}^{f(n+1)} \frac{1}{2^{i+2}}.$$

Now, let N denote 2^n. By the above inequality, it holds that

$$\frac{1}{2^N} \times \mathrm{Prob}[E_3^{n+1}|E_2^{S,n+1}] > \frac{1}{2^N} - \frac{1}{2^{f(n)+3}} \times \frac{1}{1-\frac{1}{2}} = \frac{1}{2^N} - \frac{1}{2^{f(n)+2}}.$$

Note that $f(n)+2 = N+n+3$. Therefore, the following holds.

$$\mathrm{Prob}[E_3^{n+1}|E_2^{S,n+1}] > 1 - \frac{1}{2^{n+3}}. \tag{2}$$

Now, let E_4^{n+1}, E_5^{n+1} denote the following events with respect to a randomly chosen forcing condition $T : \{z(0), z(1), \cdots, z(2^{f(n+1)}-1)\} \to \{0,1\}$.

$$E_4^{n+1} : \text{“}n+1 \leq \forall i \leq f(n+1)\ T \upharpoonright z(2^i) \in r\mathrm{GEN}_1(k,c,i)\text{”},$$

$$E_5^{n+1} : \text{“}n+1 \leq \forall i \leq f(n)\ T \upharpoonright z(2^i) \in r\mathrm{GEN}_1(k,c,i)\text{”}.$$

We review the differences among definitions of E_1^n, E_3^{n+1}, E_4^{n+1}, E_5^{n+1} and E_1^{n+1}. In Table 1, the dimension of T denotes the integer d such that the domain of T, a randomly chosen forcing condition, is of the form $\{z(0), z(1), \cdots, z(2^d-1)\}$.

Table 1. The differences among definitions of events

Event	dimension of T	range of i	
E_1^n	$f(n)$	$n+1, \cdots,$	$f(n)$
E_3^{n+1}	$f(n+1)$	$f(n)+1, \cdots,$	$f(n+1)$
E_4^{n+1}	$f(n+1)$	$n+1, \cdots,$	$f(n+1)$
E_5^{n+1}	$f(n+1)$	$n+1, \cdots,$	$f(n)$
E_1^{n+1}	$f(n+1)$	$n+2, \cdots,$	$f(n+1)$

Throughout the rest of the proof, the superscripts of E_3, E_4 and E_5 are fixed to $n+1$. Thus, we omit to write them.

Since it holds that $E_4 = E_3 \cap E_5$, the following holds.

$$\begin{aligned} &1 - \mathrm{Prob}[E_4] = \mathrm{Prob}[\mathrm{not}E_4] = \mathrm{Prob}[(\mathrm{not}E_3) \cup (\mathrm{not}E_5)] \\ \leq &(1 - \mathrm{Prob}[E_3]) + (1 - \mathrm{Prob}[E_5]). \end{aligned}$$

Therefore, it holds that $\mathrm{Prob}[E_4] \geq -1 + \mathrm{Prob}[E_3] + \mathrm{Prob}[E_5]$. In the same way, it is shown that

$$\mathrm{Prob}[E_4|E_2^{S,n+1}] \geq -1 + \mathrm{Prob}[E_3|E_2^{S,n+1}] + \mathrm{Prob}[E_5|E_2^{S,n+1}]. \quad (3)$$

Note that the only difference between E_5 and E_1^n is the domain of the randomly chosen forcing condition T: it is $\{z(0), z(1), \cdots, z(2^{f(n+1)} - 1)\}$ in the former, and it is $\{z(0), z(1), \cdots, z(2^{f(n)} - 1)\}$ in the latter. The same thing holds on the difference between $E_2^{S,n+1}$ and $E_2^{S,n}$. Hence, the following holds.

$$\mathrm{Prob}[E_5|E_2^{S,n+1}] = \mathrm{Prob}[E_1^n|E_2^{S,n}].$$

By our assumption that $R(S,n)$ is satisfied, we get the following.

$$\mathrm{Prob}[E_5|E_2^{S,n+1}] \geq \frac{1}{2}(1 + \frac{1}{2^{n+1}}). \quad (4)$$

With (2), (3) and (4), it holds that

$$\mathrm{Prob}[E_4|E_2^{S,n+1}] > -\frac{1}{2^{n+3}} + \frac{1}{2}(1 + \frac{1}{2^{n+1}}) = \frac{1}{2}(1 + \frac{1}{2^{n+2}}). \quad (5)$$

Hence, there exists a forcing condition S' such that S' extends S, the domain of S' is $\{z(0), z(1), \cdots, z(2^{n+1} - 1)\}$ and such that

$$\mathrm{Prob}[E_4|E_2^{S',n+1}] \geq \frac{1}{2}(1 + \frac{1}{2^{n+2}}). \quad (6)$$

Now, remind that E_1^{n+1} denotes the following event.

$$E_1^{n+1} : \text{“}n + 2 \leq \forall i \leq f(n+1)\ T \restriction z(2^i) \in r\mathrm{GEN}_1(k, c, i)\text{”},$$

Hence, by (6), it holds that $S' \in r\mathrm{GEN}_1(k, c, n+1)$, and that

$$\mathrm{Prob}[E_1^{n+1}|E_2^{S',n+1}] \geq \frac{1}{2}(1 + \frac{1}{2^{n+2}}).$$

In other words, $R(S', n+1)$ holds. Thus, we have shown the lemma. □

Proof of Lemma 4.1. Fix positive integers c and k such that $c \geq 2^{f(2)}$ $(= 2^7 = 128)$ and such that for every positive integers $r \geq 2$ and n, the equation (1) of Lemma 3.1 holds.

We define a primitive recursive sequence $\{S_n\}_{n\geq 1}$ of forcing conditions such that for each n the following (i)–(iii) hold.

(i) The domain of S_n is $\{z(0), z(1), \cdots, z(2^n - 1)\}$.

(ii) If $n \geq 2$ then S_n extends S_{n-1}.

(iii) The requirement $R(S_n, n)$ is satisfied, where R is the requirement introduced in Lemma 4.2.

The construction is as follows.

Let $S_1 : \{z(0), z(1)\} \to \{0\}$. (i) and (ii) hold. Since it holds that $c \geq 2^{f(2)}$, by Proposition 3.1, (iii) holds.

As the induction step, suppose that S_k are constructed for all $k < n$. By Lemma 4.2, there exists a forcing condition S_n satisfying (i)–(iii). Chose the least such S_n with respect to length-lexicographic order.

Let D be the union of $\{S_n\}_{n \geq 1}$. Then D is a primitive recursive r-Dowd oracle. □

5. Proof of Main Theorem

In this section, we prove the main theorem. The proof is more complicated than that of Lemma 4.1, but the underlying idea of the proof resembles that of Lemma 4.1. Throughout this section, for each natural number n, let

$$f(n) = 2^n + n + 1, \text{ and } g(n) = \frac{(n+1)^2}{n2^{n+1}}.$$

Proposition 5.1. *For each positive integer n, the followings hold, where N denotes 2^n.*

(1)

$$g(n) - \frac{n}{2^{n+2}} > \frac{(n+1)^2}{n(n+2)} g(n+1) = \frac{(n+1)(n+2)}{n2^{n+2}} > 0.$$

(2) If $n \geq 3$ then

$$\frac{1}{2^N} < \frac{1}{2(n+1)^2}\Big(g(n) - \frac{n}{2^{n+2}}\Big).$$

Proof. (1) This is easily verified.

(2) Let $h(x)$ be the following function, where X denotes 2^x.

$$h(x) = \frac{(x+2)2^{X-x-3}}{x(x+1)}.$$

Then, it is easy to verify that $h(3) > 1$ and that for every $x \geq 3$, the derivative $h'(x)$ is positive. Thus for every $n \geq 3$, it holds that $h(n) > 1$. Hence, the inequality in (2) holds. □

The following is a "one-step lemma" in our proof of the main theorem.

Lemma 5.1. *Fix positive integers k and ρ so that the equation (1) of Lemma 3.1 holds for every positive integers r, c, n such that $r \geq 2$ and*

$c \geq 2^{\rho}$. Recall that equation (1) is the following, where N denotes 2^n.

$$\left|\text{Func}(n) - r\text{GEN}_1(k, c, n)\right| \leq 2^{N-(n+2)}$$

For a forcing condition S and a positive integer n, we let $R'(S, n)$ denote the following requirement.

$$R'(S,n): \text{“}S \in r\text{GEN}_1(k, 2^{f(r)}, n) \text{ (for each } r \text{ such that } \rho \leq r \leq n\text{)}$$
$$\text{and } \text{Prob}[E_1^n | E_2^{S,n}] \geq \frac{1}{2}\big(1 + g(n)\big)\text{”},$$

where the events E_1^n and $E_2^{S,n}$ are defined as follows. We randomly choose a forcing condition $T : \{z(0), z(1), \cdots, z(2^{f(n)} - 1)\} \to \{0, 1\}$. Then,

$$E_1^n : \text{“}\rho \leq \forall r \leq n, \;\; n+1 \leq \forall i \leq f(n)$$
$$T \upharpoonright z(2^i) \in r\text{GEN}_1(k, 2^{f(r)}, i)\text{”},$$
$$E_2^{S,n} : \text{“ } T \text{ extends } S\text{”}.$$

Suppose that $n \geq 3$ is an integer, S is a forcing condition whose domain is $\{z(0), z(1), \cdots, z(2^n - 1)\}$ and suppose that $R'(S, n)$ holds.

Then, there exist two forcing conditions S'_j $(j = 1, 2)$ such that $S'_1 \neq S'_2$ and such that for each j, S'_j extends S, the domain of S'_j is $\{z(0), z(1), \cdots, z(2^{n+1} - 1)\}$ and that $R'(S'_j, n+1)$ is satisfied.

Proof. We recommend for the reader to consult the proof of Lemma 4.2 before reading the current proof.

Let E_3^{n+1} be the following event (with respect to a randomly chosen forcing condition $T : \{z(0), z(1), \cdots, z(2^{f(n+1)} - 1)\} \to \{0, 1\}$).

$$E_3^{n+1} : \text{“}\rho \leq \forall r \leq n+1, \;\; f(n)+1 \leq \forall i \leq f(n+1)$$
$$T \upharpoonright z(2^i) \in r\text{GEN}_1(k, 2^{f(r)}, i)\text{”}.$$

In the following, we omit the superscripts of E_3.

$$\text{Prob}[E_2^{S,n+1}] \times \text{Prob}[E_3 | E_2^{S,n+1}]$$
$$\geq \text{Prob}[E_2^{S,n+1} - \bigcup_{r=2}^{n+1} \bigcup_{i=f(n)+1}^{f(n+1)} \{T : T \upharpoonright z(2^i) \notin r\text{GEN}_1(k, 2^{f(r)}, i)\}]$$
$$\geq \text{Prob}[E_2^{S,n+1}] - \sum_{r=2}^{n+1} \sum_{i=f(n)+1}^{f(n+1)} \text{Prob}[T \upharpoonright z(2^i) \notin r\text{GEN}_1(k, 2^{f(r)}, i)].$$

Therefore, by Lemma 3.1, the following holds.

$$\mathrm{Prob}[E_2^{S,n+1}] \times \mathrm{Prob}[E_3|E_2^{S,n+1}] \geq \mathrm{Prob}[E_2^{S,n+1}] - n \times \sum_{i=f(n)+1}^{f(n+1)} \frac{1}{2^{i+2}}.$$

Therefore, the following holds.

$$\mathrm{Prob}[E_3|E_2^{S,n+1}] > 1 - \frac{n}{2^{n+3}}. \tag{7}$$

Now, let E_4^{n+1} and E_5^{n+1} denote the following events (with respect to a randomly chosen forcing condition $T : \{z(0), z(1), \cdots, z(2^{f(n+1)} - 1)\} \to \{0, 1\}$).

$$\begin{aligned} &E_4^{n+1} : \text{“}\rho \leq \forall r \leq n+1,\ \ n+1 \leq \forall i \leq f(n+1) \\ &\quad T \restriction z(2^i) \in r\mathrm{GEN}_1(k, 2^{f(r)}, i)\text{”}, \\ &E_5^{n+1} : \text{“}\rho \leq \forall r \leq n+1,\ \ n+1 \leq \forall i \leq f(n) \\ &\quad T \restriction z(2^i) \in r\mathrm{GEN}_1(k, 2^{f(r)}, i)\text{”}. \end{aligned}$$

We omit the superscripts of E_4 and E_5.

Claim 1 In the case where $r = n + 1$, the following holds. For every i such that $n+1 \leq i \leq f(n+1)$ and for every forcing condition T' of domain $\{z(0), z(1), \cdots, z(2^i - 1)\}$, it holds that $T' \in r\mathrm{GEN}_1(k, 2^{f(r)}, i)$.

Proof of Claim 1: Suppose that $r = n + 1$ and that T' is as above. For every i such that $n + 1 \leq i \leq f(n + 1)$, it holds that $2^{f(r)} = 2^{f(n+1)} \geq 2^i$. Hence, by Proposition 3.1, it holds that $T' \in r\mathrm{GEN}_1(k, 2^{f(r)}, i)$. Q.E.D. (Claim 1)

By Claim 1, the ranges of r in the definitions of E_3, E_4 and E_5 may be replaced by “$\rho \leq r \leq n$”.

In the same way as in the proof of Lemma 4.2, it is shown that

$$\mathrm{Prob}[E_4|E_2^{S,n+1}] \geq -1 + \mathrm{Prob}[E_3|E_2^{S,n+1}] + \mathrm{Prob}[E_5|E_2^{S,n+1}]. \tag{8}$$

With Claim 1, it is shown that the only difference between E_5 and E_1^n is the domain of the randomly chosen forcing condition T: it is $\{z(0), z(1), \cdots, z(2^{f(n+1)} - 1)\}$ in the former, and it is $\{z(0), z(1), \cdots, z(2^{f(n)} - 1)\}$ in the latter. The same thing holds on the difference between $E_2^{S,n+1}$ and $E_2^{S,n}$. Hence, the following holds.

$$\mathrm{Prob}[E_5|E_2^{S,n+1}] = \mathrm{Prob}[E_1^n|E_2^{S,n}].$$

Therefore, by our assumption that R' is satisfied, we get the following.

$$\mathrm{Prob}[E_5|E_2^{S,n+1}] \geq \frac{1}{2}(1 + g(n)). \tag{9}$$

With (7), (8) and (9), it holds that

$$\text{Prob}[E_4|E_2^{S,n+1}] > -\frac{n}{2^{n+3}} + \frac{1}{2}\big(1+g(n)\big) = \frac{1}{2}\big(1+g(n)-\frac{n}{2^{n+2}}\big). \quad (10)$$

Now, assume for a contradiction that there exists at most one forcing condition S' such that S' extends S, the domain of S' is $\{z(0), z(1), \cdots, z(2^{n+1}-1)\}$ and that

$$\text{Prob}[E_4|E_2^{S',n+1}] \geq \frac{1}{2}\big(1+g(n+1)\big). \quad (11)$$

Claim 2 Under the above assumption, the following holds, where N denotes 2^n.

$$\text{Prob}[E_4|E_2^{S,n+1}] < \frac{1}{2^N} + \frac{1}{2}\big(1+g(n+1)\big).$$

Proof of Claim 2: If no S' satisfies (11) then the above inequality is obvious.

Otherwise, by our assumption, just one S' satisfies (11). Then, for a given forcing condition that extends S, we investigate two cases. One case is that a given condition is the S', and the other case is otherwise. By investigating the sum of conditional probabilities with respect to the above two cases, we get the above inequality. Q.E.D. (Claim 2)

By Claim 2 and Proposition 5.1 (1), the following holds.

$$\text{Prob}[E_4|E_2^{S,n+1}] < \frac{1}{2^N} + \frac{1}{2}\big[1+\big(g(n)-\frac{n}{2^{n+2}}\big)\frac{n(n+2)}{(n+1)^2}\big]. \quad (12)$$

By (10) and (12), the following holds.

$$\begin{aligned}\frac{1}{2^N} &> \frac{1}{2}\big(1+g(n)-\frac{n}{2^{n+2}}\big) - \frac{1}{2}\big[1+\big(g(n)-\frac{n}{2^{n+2}}\big)\frac{n(n+2)}{(n+1)^2}\big] \\ &= \frac{1}{2}\big(g(n)-\frac{n}{2^{n+2}}\big)\frac{1}{(n+1)^2}.\end{aligned}$$

By Proposition 5.1 (2), this is a contradiction. Hence, there exist at least two mutually distinct forcing conditions S'_j $(j=1,2)$ such that for each j, S'_j extends S, the domain of S'_j is $\{z(0), z(1), \cdots, z(2^{n+1}-1)\}$ and that

$$\text{Prob}[E_4|E_2^{S'_j,n+1}] \geq \frac{1}{2}\big(1+g(n+1)\big).$$

Therefore, for each $j=1,2$, it holds that

$$\rho \leq \forall r \leq n+1 \;\; S'_j \in r\text{GEN}_1(k, 2^{f(r)}, n+1)$$

and it holds that

$$\text{Prob}[E_1^{n+1}|E_2^{S'_j,n+1}] \geq \frac{1}{2}\big(1+g(n+1)\big).$$

Hence, for each $j = 1, 2$, the requirement $R'(S'_j, n+1)$ is satisfied. Thus we have shown Lemma 5.1. □

Theorem 5.1. *(Main theorem)*

(1) There exists a primitive recursive oracle D such that for every positive integer r, D is r-Dowd.

(2) Every Turing degree contains an oracle D such that for every positive integer r, D is r-Dowd.

Proof. With Lemma 5.1, (1) is shown in a way similar to the proof of Lemma 4.1. Then, (2) of the main theorem is shown by constructing a binary tree, in the same way as the proof of Proposition 1.1 (2) in Introduction section. □

Acknowledgments

The authors would like to thank K. Ambos-Spies for his comments on our talk at the 11th Asian Logic Conference.

References

1. Suzuki, T. and Kumabe, M.: Weak randomness, genericity and Boolean decision trees. *The proceedings of the 10th Asian logic conference* (Kobe university, Kobe, Japan, September 1–6, 2008), World Scientific, pp. 322–344 (2010).
2. Suzuki, T.: Degrees of Dowd-type generic oracles. *Inf. Comput.*, **176**, pp. 66–87 (2002).
3. Dowd, M.: Generic oracles, uniform machines, and codes. *Inf. Comput.*, **96**, pp. 65–76 (1992).
4. Suzuki, T.: *Computational complexity of Boolean formulas with query symbols.* Doctoral dissertation, Institute of Mathematics, University of Tsukuba, Tsukuba-City, Japan (1999).
`http://www.ac.auone-net.jp/~bellp/toshio_suzuki_phd.pdf`
5. Suzuki, T.: Forcing complexity: minimum sizes of forcing conditions. *Notre Dame J. Formal Logic*, **42**, pp. 117–120 (2001).
6. Kumabe, M., Suzuki, T. and Yamazaki, T.: Does truth-table of linear norm reduce the one-query tautologies to a random oracle? *Arch. Math. Logic*, **47**, pp. 159–180 (2008).
7. Lutz, J. H.: Category and measure in complexity classes. *SIAM J. Comput.*, **19**, pp. 1100–1131 (1990).
8. Lutz, J. H.: Almost everywhere high nonuniform complexity. *J. Comput. System Sci.*, **44**, pp. 220–258 (1992).
9. Martin-Löf, P.: The definition of random sequences. *Information and Control*, **9**, pp. 602–619 (1966).

10. Schnorr, C. P.: Zufälligkeit und Warscheinlichkeit. In: *Lecture notes in mathematics 218*, Springer, New York, 1971.
11. Downey, R., Hirschfeldt, D. R., Nies, A. and Terwijn, S. A.: Calibrating randomness. *Bull. Symb. Log.*, **12**, pp. 411–491 (2006).
12. Ambos-Spies, K., Terwijn, S. A. and Zheng, X.: Resource bounded randomness and weakly complete problems. *Theoret. Comput. Sci.*, **172**, pp. 195–207 (1997).

MODELS OF LONG SENTENCES I

Gerald E. Sacks
Harvard University
sacks@math.harvard.edu

For Professor Gert Müller (in memoriam)
and
For Professor Chi Tat Chong on his 60th birthday

Let $\mathcal{L}$ be a countable first order language. Let A be a Σ_1 admissible set such that $\mathcal{L} \in A$ and the cardinality of A is ω_1. Let $T \subseteq \mathcal{L}_{A,\omega}$ be a set of sentences such that $< A, T >$ is Σ_1 admissible. T is **consistent** iff no contradiction can be derived from T via a deduction in A. T is **complete** iff for each sentence $\mathcal{F} \in \mathcal{L}_{A,\omega}$, either $\mathcal{F} \in T$ or $(\neg\mathcal{F}) \in T$. An ***n*-type** p of T is a consistent, complete set of formulas of arity $\leq n$ such that $< A, p >$ is Σ_1 admissible; ST is the set of all types of T.

Say T is **degenerate** iff T has a countable, ω-homogeneous model that realizes every type in ST.

A typical instance of **type-completeness** is: if $\exists y \mathcal{F}(\overline{x}, y) \in p(\overline{x}) \in ST$, then there is a $q(\overline{x}, y) \in ST$ such that $p(\overline{x}) \subseteq q(\overline{x}, y)$ and $\mathcal{F}(\overline{x}, y) \in q(\overline{x}, y)$.

T is **type-admissible** iff $< A, \overline{p} >$ is Σ_1 admissible for each coherent pair $\overline{p}$ of types.

Mild stability (Section 6) implies type-completeness and type-admissibilty.

Main Result: If T is consistent, complete, type-complete, type-admissible, and not degenerate, then T has a model of size ω_1.

Contents

My thanks to Professor Douglas Cenzer and the Mathematics Department of the University of Florida for the opportunity to speak on matters related to this paper during Model Theory Week of the Special Year in Logic, 2006-2007.

1. Introduction

This paper stands on its own four feet[1] but also serves as an introduction to the arguments of its sequel [5], where longer sentences, larger models and more complex countable approximations are studied. The proof of the **Main Result** below was inspired by the work of Barwise [1] and his students on admissible sets and their application to model theory [4]. Jensen's proof [2] of the gap-2 conjecture in L plays a part, behind the scenes in this paper, but on stage in its sequel [5].

Let $\mathcal{L}$ be a countable first order language. Recall that $\mathcal{L}_{\infty,\omega}$ is an extension of first order logic that allows arbitrary conjunctions and disjunctions of formulas subject to the restriction that a formula can contain only finitely many free variables. On the other hand a formula can mention arbitrarily many individual constants.

Recall that a set A is $\mathbf{\Sigma_1}$ **admissible** iff A is transitive, closed under pairing and unary unions, and satisfies Δ_0 separation and Δ_0 collection (or bounding); Σ_1 admissibility implies Δ_1 separation and Σ_1 collection.

From now on assume A is a Σ_1 admissible set such that $\mathcal{L} \in A$ and the cardinality of A is ω_1.

(Note: the uncountability of A does not imply $\omega_1 \subseteq A$.) For any $Z \subseteq A$, define

$$A[Z] = \cup\{L(\alpha, tc(\{a\}); Z) \mid a \in A\}, \tag{1.1}$$

where α is the least ordinal not in A, and $L(\alpha, tc(\{a\}); Z)$ is the result of iterating first order definability, with $x \in Z$ as an additional Δ_0 formula, through the ordinals less than α, and with $tc(\{a\})$ as the starting set (tc is transitive closure). The structure $< A[Z], Z >$ is said to be $\mathbf{\Sigma_1}$ **admissible** iff $A[Z]$ is Σ_1 admissible with $x \in Z$ as an additional Δ_0 formula. Define $\mathcal{L}_{A,\omega}$ to be be the restriction of $\mathcal{L}_{\infty,\omega}$ to formulas with standard codes in A.

[1]With thanks to Theodore Slaman.

Assume $T \subseteq \mathcal{L}_{A,\omega}$ is a set of sentences such that $< A[T], T >$ is Σ_1 admissible.

Definition 1. *Suppose $Z \subseteq A$; Z is **amenable** iff $(Z \cap b) \in A$ for every $b \in A$.*

Remark 2. *Recall $A[Z] = A$ iff Z is amenable.*

Assume T is amenable. Thus $< A, T >$ is Σ_1 admissible.

Definition 3. *T is **consistent** iff no contradiction can be derived from T using the axioms and rules of $\mathcal{L}_{\infty,\omega}$ via a deduction that belongs to A. (The axioms and rules of $\mathcal{L}_{\infty,\omega}$ extend first order logic primarily by adding an infinitary conjunction rule: if $\mathcal{F}_i$ is deducible for each $i \in I$, then $\wedge\{\mathcal{F}_i \mid i \in I\}$ is deducible.) As a rule, in all that follows, a set $Z \subseteq \mathcal{L}_{A,\omega}$ will be said to be **consistent** iff $< A, Z >$ is Σ_1 admissible (this implies Z is amenable), and no deduction in A from Z yields a contradiction. (Also the set of free variables occurring in Z is finite.)*

Remark 4. *Let $Z \subseteq \mathcal{L}_{A,\omega}$ be as in **Definition 3**. By Barwise Z is consistent in the strongest syntactical sense: no deduction in V (the class of all sets) from Z using the axioms and rules of $\mathcal{L}_{\infty,\omega}$ yields a contradiction. (Recall that Barwise completeness holds for Σ_1 admissible structures of any size.)*

Definition 5. *T is **complete** iff for each sentence $\mathcal{F} \in \mathcal{L}_{A,\omega}$, $\mathcal{F} \in T$ or $(\neg\mathcal{F}) \in T$.*

Definition 6. *Let $\overline{x}$ denote a sequence $x_1, ...x_n$ of n distinct free variables. A formula has arity n iff the number of distinct free variables occurring in it is n. An **n-type** $p(\overline{x})$ of T is a set of formulas whose free variables occur in $\overline{x}$ and such that:*
(i) $p(\overline{x}) \subseteq \mathcal{L}_{A,\omega}$ and $p(\overline{x})$ is amenable;
(ii) For each $\mathcal{G}(\overline{x}) \in \mathcal{L}_{A,\omega}$ of arity $\leq n$, either $\mathcal{G}(\overline{x}) \in p(\overline{x})$ or $(\neg\mathcal{G}(\overline{x})) \in p(\overline{x})$;
(iii) The structure $< A, p(\overline{x}) >$ is Σ_1 admissible;
(iv) $T \subseteq p(\overline{x})$ and $p(\overline{x})$ is consistent.

Definition 7. *ST is the set of all n-types of T for all $n > 0$.*

Proposition 8. *Suppose $\vee\{\mathcal{G}_i(\overline{x}) \mid i \in I\} \in p(\overline{x})$.Then for some $i_0 \in I$, $\mathcal{G}_{i_0}(\overline{x}) \in p(\overline{x})$.*

Proof. Suppose not. Then $(\neg\mathcal{G}_i(\overline{x})) \in p(\overline{x})$ for all $i \in I$. By clause (iii) of Definition 6,

$$\wedge\{\neg\mathcal{G}_i(\overline{x}) \mid i \in I\} \tag{1.2}$$

is deducible from $p(\overline{x})$ via a deduction in A. But then $p(\overline{x})$ is inconsistent. ■

A type is presented as a set $p(\overline{x})$ of formulas whose free variables belong to $\overline{x}$. The choice of $\overline{x}$ matters.

For $\overline{v}$ a subsequence of $\overline{x}$ ($\overline{v} \subseteq \overline{x}$), define

$$p(\overline{v}) = \{\mathcal{F}(\overline{v}) \mid \mathcal{F}(\overline{v}) \in p(\overline{x})\}. \tag{1.3}$$

Definition 9. *Suppose* $p_1(\overline{x^1}), p_2(\overline{x^2}) \in ST$. *Let* $\overline{v}$ $(= \overline{x^1} \cap \overline{x^2})$ *be the sequence of variables common to* $\overline{x^1}$ *and* $\overline{x^2}$. *The pair,* $p_1(\overline{x^1}), p_2(\overline{x^2})$, *is said to be* ***coherent*** *iff* $p_1(\overline{v}) = p_2(\overline{v})$.

Definition 10. T *is* ***type-admissible*** *iff* $< A, \overline{p} >$ *is* Σ_1 *admissible for each coherent pair* $\overline{p}$ *of types in* ST.

Type-admissibility is needed for the amalgamation of types during the construction of the model $\mathcal{B}_{\omega_1}$ in the proof of the Main Result. In some situations it can be dropped, cf. Subsections 5.2 and 5.3.

Proposition 11. *Suppose* T *is amenable, consistent, complete and type-admissible. If* $p_1(\overline{x^1}), p_2(\overline{x^2})$ *is a coherent pair of types, then* $p_1(\overline{x^1}) \cup p_2(\overline{x^2})$ *is consistent.*

Proof. Suppose not. Then there is a deduction in A of a contradiction from $\mathcal{F}_1(\overline{x^1}) \wedge \mathcal{F}_2(\overline{x^2})$ for some $\mathcal{F}_i(\overline{x^i}) \in p_i(\overline{x^i})$ $(i = 1, 2)$. Let $\overline{v}$ be $\overline{x^1} \cap \overline{x^2}$, and $\overline{u^i}$ be $\overline{x^i} - \overline{v}$. Then

$$\exists\overline{u^1}\mathcal{F}_1(\overline{u^1}, \overline{v}) \wedge \exists\overline{u^2}\mathcal{F}_2(\overline{u^2}, \overline{v}) \tag{1.4}$$

yields a contradiction. But the coherence of p_1 and p_2 implies formula (1.4) belongs to $p_1(\overline{x^1})$. ■

Notation 12. Let $\overline{x^1 \cup x^2}$ denote a sequence of distinct free variables, every one of which occurs in $\overline{x^1}$ or $\overline{x^2}$.

Definition 13. T *is* ***type-complete*** *iff:*
Suppose $p_1(\overline{x^1}), p_2(\overline{x^2}) \in ST$ *and*

$$\{\exists y\mathcal{G}(\overline{x^1 \cup x^2}, y)\} \cup p_1(\overline{x^1}) \cup p_2(\overline{x^2}) \tag{1.5}$$

is consistent. Then there exists an $r(\overline{x^1 \cup x^2}, y) \in ST$ *such that* $\mathcal{G}(\overline{x^1 \cup x^2}, y) \in r(\overline{x^1 \cup x^2}, y)$ *and*

$$p_1(\overline{x^1}), p_2(\overline{x^2}) \subseteq r(\overline{x^1 \cup x^2}, y). \tag{1.6}$$

Every consistent, complete theory contained in a countable fragment of $\mathcal{L}_{\omega_1,\omega}$ is type-complete. In the uncountable case a type-complete theory has advantages similar to those of an atomic theory, cf. Subsection 5.2.

Proposition 14. *Suppose* T *is amenable, consistent, complete, type-admissible and type-complete. If* $p_1(\overline{x^1}) \cup p_2(\overline{x^2})$ *is consistent, then there exists an* $r(\overline{x^1 \cup x^2}) \in ST$ *such that* $p_1(\overline{x^1}), p_2(\overline{x^2}) \subseteq r(\overline{x^1 \cup x^2})$.

Proof. An instance of type-completeness with

$$[(\overline{x^1 \cup x^2} = \overline{x^1 \cup x^2})] \wedge (y = y) \tag{1.7}$$

as $\mathcal{G}(\overline{x^1 \cup x^2}, y)$. ■

Definition 15. T *is* ***degenerate*** *iff* T *has a countable,* ω*-homogeneous model that realizes every type in* ST.

Main Result (MR). If T is amenable, consistent, complete, type-complete, type-admissible, and not degenerate, then T has a model $\mathcal{B}$ of cardinality ω_1.

MR+. In addition: if $X \subseteq ST$ and $card(X) = \omega_1$, then $\mathcal{B}$ can be made to realize all the types in X.

The proof of MR given below, after a minor adjustment in Subsection 5.2, yields an atomic model when T is atomic; in that case the assumptions of amenability, type-completeness and type-admissibility can be dropped.

Let $\mathcal{B}_{\omega_1}$ be the model whose existence is claimed in the main result. The structure $\mathcal{B}_{\omega_1}$ is a Henkin-style model that is the limit of a chain of countable partial Henkin models, $\mathcal{B}_\gamma$ $(\gamma < \omega_1)$. $\mathcal{B}_\gamma$ is said to be "partial," because it is the result of a Henkin-type construction $\mathbf{C}_\gamma$ of length ω carried out on a countable set of formulas that may lack the subformula property. The construction $\mathbf{C}_\gamma$ builds a partial model of T_γ, the intersection of T with a countable Σ_1 hull H_γ. The theory T_γ may have a sentence $\mathcal{G}$ of uncountable length; consequently $\mathcal{G}$ will be declared true by $\mathbf{C}_\gamma$, but if $\mathcal{G}$ has an existential subsentence not in H_γ, then that subsentence will not be assigned an existential witness by $\mathbf{C}_\gamma$. On the other hand each n-tuple of Henkin constants of $\mathcal{B}_\gamma$ is assigned to some n-type of T. The hull H_γ ensures that T_γ is a good approximation of T, an approximation that gets better as γ increases. The embeddability of $\mathcal{B}_\gamma$ in $\mathcal{B}_{\gamma+1}$ results from a property of $\mathcal{B}_{\gamma+1}$ akin to ω-saturation.

2. Σ_1 Substructures

Let H be a Σ_1 substructure of the universe V ($H \preceq_1 V$) such that the cardinality of H is ω_1, $A \subseteq H$, and $A, T, ST \in H$. Define

$$ST_H = ST \cap H. \tag{2.1}$$

Proposition 16. *If ST_H is countable, then $ST_H = ST$.*

Proof. Suppose ST is uncountable. In H there is an one-one map of A into ST. But then ST_H is uncountable since $A \subseteq H$. Thus ST is countable. In H there is a map of ω onto ST, so $ST \subseteq H$. ■

Proposition 17. *There exists a chain H_γ ($\gamma < \omega_1$) of countable structures such that*

$$A, T \in H_0, \tag{2.2}$$

$$H_\gamma \preceq_1 H_{\gamma+1} \quad (\gamma < \omega_1), \tag{2.3}$$

$$H_\lambda = \cup\{H_\gamma \mid \gamma < \lambda\} \quad (\textit{limit } \lambda < \omega_1), \tag{2.4}$$

$$H = \cup\{H_\gamma \mid \gamma < \omega_1\}. \tag{2.5}$$

3. Akin to ω-Saturation

For $\gamma < \omega_1$, define

$$\mathcal{L}_{\gamma,\omega} = \mathcal{L}_{A,\omega} \cap H_\gamma, \tag{3.1}$$

$$T_\gamma = T \cap H_\gamma, \tag{3.2}$$

$$ST_\gamma = ST_H \cap H_\gamma. \tag{3.3}$$

Then $\mathcal{L}_{A,\omega}$ is $\cup\{\mathcal{L}_{\gamma,\omega} \mid \gamma < \omega_1\}$, T is $\cup\{T_\gamma \mid \gamma < \omega_1\}$, and ST_H is $\cup\{ST_\gamma \mid \gamma < \omega_1\}$.

Fix $\gamma < \omega_1$. From an intuitive point of view, the structure $\mathcal{B}_\gamma$ is the result of building a countable ω-homogeneous structure that realizes all the types in ST_γ and no others. Note that $ST_\gamma \subseteq ST$.

Construction of $\mathcal{B}_\gamma$. Fix $s < \omega$. Prior to stage s of the construction, a sequence $\overline{c^{\gamma,s}}$ of distinct individual constants $c_1^\gamma, ..., c_s^\gamma$ was developed ($\overline{c^{\gamma,0}}$ is null). A type $r_{\gamma,s}(\overline{x^{\gamma,s}}) \in ST_\gamma$ was assigned to $\overline{c^{\gamma,s}}$; thus $\overline{x^{\gamma,s}}$ denotes $x_1^\gamma, ..., x_s^\gamma$, and the result of the construction prior to stage s is the set of sentences $r_{\gamma,s}(\overline{c^{\gamma,s}})$ ($r_{\gamma,0}$ is T). Suppose $\overline{v}$ is a subsequence of $\overline{x^{\gamma,s}}$ and $\overline{d}$ is a subsequence of $\overline{c^{\gamma,s}}$ that realizes $r_{\gamma,s}(\overline{v})$; i.e. $r_{\gamma,s}(\overline{d}) \subseteq r_{\gamma,s}(\overline{c^{\gamma,s}})$. Let y be a variable not occurring in $\overline{x^{\gamma,s}}$.

Case I (existential witnesses). Suppose $\mathcal{G}(\overline{v}, y) \in \mathcal{L}_{\gamma,\omega}$ and

$$\exists y \mathcal{G}(\overline{d}, y) \in r_{\gamma,s}(\overline{d}). \tag{3.4}$$

Then $\{\exists y \mathcal{G}(\overline{v}, y)\} \cup r_{\gamma,s}(\overline{v})$ is consistent. By the type-completeness of T, there is an $r(\overline{v}, y) \in ST$ such that

$$\mathcal{G}(\overline{v}, y) \in r(\overline{v}, y) \text{ and } r_{\gamma,s}(\overline{v}) \subseteq r(\overline{v}, y). \tag{3.5}$$

By Propositions 11 and 14 there is an $r'(\overline{x^{\gamma,s}}, y) \in ST$ such that $r_{\gamma,s}(\overline{x^{\gamma,s}}), r(\overline{v}, y) \subseteq r'(\overline{x^{\gamma,s}}, y)$. And $r'(\overline{x^{\gamma,s}}, y)$ can be taken from ST_γ, since $H_\gamma \preccurlyeq_1 H \preccurlyeq_1 V$.

Let e be an individual constant not occurring in $\overline{c^{\gamma,s}}$ or in $\mathcal{L}$ or in $\mathcal{B}_\delta$ for any $\delta < \gamma$ Define:

$$c^\gamma_{s+1} = e; \tag{3.6}$$

$$\overline{c^{\gamma,s+1}} \text{ is the sequence } \overline{c^{\gamma,s}}, c^\gamma_{s+1}; \tag{3.7}$$

$$r_{\gamma,s+1}(\overline{x^{\gamma,s+1}}) = r'(\overline{x^{\gamma,s}}, y). \tag{3.8}$$

The result of the construction at the end of stage s is the set of sentences $r_{\gamma,s+1}(\overline{c^{\gamma,s+1}})$. And $\mathcal{G}(\overline{d}, c^\gamma_{s+1}) \in r_{\gamma,s+1}(\overline{c^{\gamma,s+1}})$.

Case 2 (homogeneity and universality). Suppose $q(\overline{v}, y) \in ST_\gamma$ and $r_{\gamma,s}(\overline{v}) \subseteq q(\overline{v}, y)$. As in Case 1, the type-completeness of T implies there is an $r'(\overline{x^{\gamma,s}}, y) \in ST_\gamma$ such that

$$r_{\gamma,s}(\overline{x^{\gamma,s}}), q(\overline{v}, y) \subseteq r'(\overline{x^{\gamma,s}}, y). \tag{3.9}$$

Let e and $r_{\gamma,s+1}(\overline{x^{\gamma,s+1}})$ be as in Case 1. Then $\overline{d}, e$ realizes $q(\overline{v}, y)$ and the result of the construction at the end of stage s is $r_{\gamma,s+1}(\overline{c^{\gamma,s}}, e)$.

Define

$$\mathcal{B}_\gamma = \cup\{r_{\gamma,s}(\overline{c^{\gamma,s}}) \mid s < \omega\} \tag{3.10}$$

$$c^\gamma \text{ is } c^\gamma_1, c^\gamma_2, \ldots c^\gamma_s, \ldots (s < \omega). \tag{3.11}$$

The universe of $\mathcal{B}_\gamma$ is c^γ. Let $\overline{d}$ denote an n-tuple of $\mathcal{B}_\gamma$ ($n \geq 0$).

Proposition 18. *(i) $\mathcal{B}_\gamma$ is a set of sentences of the form $\mathcal{F}(\overline{d})$, where $\mathcal{F}(\overline{x})$ is a formula of $\mathcal{L}_{\gamma,\omega}$ and $\overline{d}$ is a subsequence of c^γ. Every sentence of this form, or its negation, belongs to $\mathcal{B}_\gamma$.*

(ii) Each $\overline{d}$ *has been assigned a type* $p(\overline{x}) \in ST_\gamma$, *and* $\mathcal{B}_\gamma$ *is the union of all such* $p(\overline{d})$*'s.*
(iii) Suppose $\exists y \mathcal{G}(\overline{x}, y) \in \mathcal{L}_{\gamma,\omega}$ *and* $\exists y \mathcal{G}(\overline{d}, y) \in \mathcal{B}_\gamma$; *then* $\mathcal{G}(\overline{d}, c_s^\gamma) \in \mathcal{B}_\gamma$ *for some* s.
(iv) Suppose $p(\overline{x})$ *has been assigned to* $\overline{d}$ *and* $p(\overline{x}) \subseteq q(\overline{x}, y) \in ST_\gamma$; *then* $q(\overline{x}, y)$ *has been assigned to* $\overline{d}, e$ *for some* $e \in c^\gamma$ *(*$p(\overline{x})$ *can be null).*

Remark 19. *In the light of Proposition 18 it is reasonable to say:* $\mathcal{B}_\gamma$ *is a* ω***-homogeneous*** *model of* T_γ *that realizes all the types in* ST_γ*; keep in mind that* $ST_\gamma \subseteq ST$.

4. Proof of the Main Result

An injective, type preserving map

$$m_{\gamma,\gamma+1} : c^\gamma \to c^{\gamma+1} \tag{4.1}$$

is defined by recursion on $s < \omega$. Prior to stage s of the recursion, a sequence

$$\overline{b^{\gamma+1,s}} = b_1^{\gamma+1}, ..., b_s^{\gamma+1} \tag{4.2}$$

was developed so that $r_{\gamma,s}(\overline{b^{\gamma+1,s}}) \subseteq \mathcal{B}_{\gamma+1}$. ($\overline{b^{\gamma+1,0}}$ is null.) By Proposition 18(iv),

$$r_{\gamma,s+1}(\overline{b^{\gamma+1,s}}, e) \subseteq \mathcal{B}_{\gamma+1} \tag{4.3}$$

for some $e \in c^{\gamma+1}$. Define $m_{\gamma,\gamma+1}(c_{s+1}^\gamma) = e = b_{s+1}^{\gamma+1}$.

The map $m_{\gamma,\gamma+1} : \mathcal{B}_\gamma \to \mathcal{B}_{\gamma+1}$ is defined by

$$m_{\gamma,\gamma+1}(\mathcal{F}(\overline{d})) = \mathcal{F}(m_{\gamma,\gamma+1}(\overline{d})). \tag{4.4}$$

Thus the type assigned to $\overline{d}$ in $\mathcal{B}_\gamma$ equals the type assigned to $m_{\gamma,\gamma+1}(\overline{d})$ in $\mathcal{B}_{\gamma+1}$.

A direct system, $\{\mathcal{B}_\gamma, m_{\gamma,\delta} \mid \gamma < \delta < \omega\}$ is defined by recursion on ω_1.

Define $m_{\gamma,\delta+1} = m_{\delta,\delta+1} m_{\gamma,\delta}$. If λ is a countable limit, then let $\mathcal{B}'_\lambda, m'_{\gamma,\lambda}$ $(\gamma < \lambda)$ be the direct limit of $\{\mathcal{B}_\gamma, m_{\gamma,\delta} \mid \gamma < \delta < \lambda\}$. Recall the terminology suggested in Remark 19. By Proposition 18, both $\mathcal{B}'_\lambda$ and $\mathcal{B}_\lambda$ are ω-homogeneous and realize the same set of types, ST_λ. A back-and-forth argument produces an isomorphism

$$i_\lambda : \mathcal{B}'_\lambda \to \mathcal{B}_\lambda. \tag{4.5}$$

Define $m_{\gamma,\lambda} = i_\lambda m'_{\gamma,\lambda}$.

Definition 20. *$\mathcal{B}_{\omega_1}$ is the direct limit of $\{\mathcal{B}_\gamma, m_{\gamma\delta} \mid \gamma < \delta < \lambda < \omega_1\}$.*

Lemma 21. *(i) $\mathcal{B}_{\omega_1}$ is finitarily consistent; i.e. no deduction from $\mathcal{B}_{\omega_1}$ of finite length yields a contradiction.*
(ii) If a disjunction belongs to $\mathcal{B}_{\omega_1}$, then some term of the disjunction belongs to $\mathcal{B}_{\omega_1}$.
(iii) $\mathcal{B}_{\omega_1}$ has existential witnesses.
(iv) (completeness) Suppose $\mathcal{F}(\overline{x})$ belongs to $\mathcal{L}_{A,\omega}$ and $\overline{d}$ is a sequence of individual constants of $\mathcal{B}_{\omega_1}$, constants of the form $m_{\gamma,\omega_1}(c_s^\gamma)$; then $\mathcal{F}(\overline{d})$, or its negation, belongs to $\mathcal{B}_{\omega_1}$.
(v) Let $\mathcal{B}_{\omega_1}$ ambiguously denote the Henkin-style model defined by the sentences of $\mathcal{B}_{\omega_1}$. Then $\mathcal{B}_{\omega_1}$ is ω-homogeneous and the set of types realized in $\mathcal{B}_{\omega_1}$ is ST_H.
(vi) $\mathcal{B}_{\omega_1}$ has cardinality ω_1.

Proof. Assertions (i)-(iv) follow from Proposition 18. A Henkin-style model is determined by $\mathcal{B}_{\omega_1}$ as in first order logic thanks to (i)-(iv). Suppose $\mathcal{B}_{\omega_1}$ is countable. Hence $\mathcal{B}_{\omega_1} = \mathcal{B}_{\gamma_0}$ for some countable γ_0. Then ST_H is countable by Lemma 21(iii), hence $ST_H = ST$ by Proposition 16, and so T is degenerate. ■

MR+. In addition: if $X \subseteq ST$ and $card(X) = \omega_1$, then $\mathcal{B}$ can be made to realize all the types in X.

Proof. A slight modification of Sections 2 and 3. Let H^X be a Σ_1 substructure of V such that $card(H^X) = \omega_1$; $X, A \subseteq H^X$; and $X, A, T, ST \in H^X$. Define

$$ST_H^X = (ST \cap H^X). \tag{4.6}$$

There exists a chain H_γ^X $(\gamma < \omega_1)$ of countable structures such that

$$X, A, T \in H_0^X, \tag{4.7}$$

$$H_\gamma^X \preceq_1 H_{\gamma+1}^X, \tag{4.8}$$

$$H_\lambda^X = \cup\{H_\gamma^X \mid \gamma < \lambda\} \ (limit\ \lambda), \tag{4.9}$$

$$H^X = \cup\{H_\gamma^X \mid \gamma < \omega_1\}. \tag{4.10}$$

Define $ST_\gamma^X = ST_H^X \cap H_\gamma^X$ $(\gamma < \omega_1)$. Then $ST_H^X = \cup\{ST_\gamma^X \mid \gamma < \omega_1\}$. Now proceed as in Section 3. The set of types realized in $\mathcal{B}_\gamma$ will be ST_γ^X. ■

5. Extensions of MR and MR+

5.1. *The number of models*

Corollary 22. *Assume T is amenable, consistent, complete, type-complete, type-admissible, and not degenerate. If $card(ST) > \omega_1$, then the number of models of T of cardinality ω_1 is at least $card(ST)$.*

Proof. Let $\{p_\gamma \mid \gamma < card(ST)\}$ be an enumeration of ST. Let $\mathcal{C}^\gamma$ be a model of T of cardinality ω_1 that realizes p_γ. Then the number μ of models of T (up to isomorphism) in the sequence $\{\mathcal{C}^\gamma \mid \gamma < card(ST)\}$ is at least $card(ST)$; otherwise $card(ST) \leq \mu \times \omega_1 < card(ST)$. ∎

5.2. *Atomic theories*

In this subsection, as in the Introduction, $\mathcal{L}$ is a countable first order language, A is a Σ_1 admissible set of cardinality ω_1, $T \subseteq \mathcal{L}_{A,\omega}$, and $< A, T >$ is Σ_1 admissible; T is said to be **atomic** if for each consistent formula $\mathcal{F}(\overline{x})$,there is an atom $\mathcal{G}(\overline{x})$ of T such that $T \vdash [\mathcal{G}(\overline{x}) \rightarrow \mathcal{F}(\overline{x})]$.

Corollary 23. *Assume T is consistent, complete and atomic. Suppose T does not have a countable atomic model. Then T has an atomic model of cardinality ω_1.*

Proof. A small modification of Sections 2, 3 and 4. Define aT to be the set of atoms of T. (There are no repetitions in aT; each atom of T has just one formula representing it in aT.) Replace ST by aT in the definition of H in Section 2. Define H_γ as in Section 2. Let $aT_\gamma = aT \cap H_\gamma$. Then $aT = \cup\{aT_\gamma \mid \gamma < \omega_1\}$. Now proceed as in Section 3. In both cases of the construction of $\mathcal{B}_\gamma$, the type r' is an atom. As in Remark 19, it is reasonable to say $\mathcal{B}_\gamma$ is ω-homogeneous and the set of types realized in $\mathcal{B}_\gamma$ is aT_γ. If $\mathcal{B}_{\omega_1}$ were countable, then T would have a countable atomic model. ∎

5.3. $\mathcal{L}_{\omega_1,\omega}$

Let HC be the set of hereditarily countable sets. Let $\mathcal{L}$ be countable and $\in HC$. Then $\mathcal{L}_{\omega_1,\omega}$ is $\mathcal{L}_{HC,\omega}$. For any $Z \subseteq HC$: Z is amenable and $< HC, Z >$ is Σ_1 admissible.

Corollary 24. *Assume the Continuum Hypothesis. If $T \subseteq \mathcal{L}_{\omega_1,\omega}$ is consistent, complete, type-complete and not degenerate, then T has a model of cardinality ω_1.*

6. Stability, Type-Completeness and Type-Admissibility

This section outlines some of the points made in [5]. Suppose $\mathcal{L}$ is a countable first order language, A is a Σ_1 admissible set, and $T \subseteq \mathcal{L}_{A,\omega}$ is a set of sentences such that $< A, T >$ is Σ_1 admissible and T is consistent and complete as in Section 1. Note that no assumption is made about the cardinality of A.

Does T have a model? A seemingly simpler question is: Does T have any types? The latter can be answered with the help of a suitable notion of stability. Call T **mildly stable** if ST', the set of types of T', is countable whenever T' is a countable subtheory of T.

A sketch of a proof that $ST \neq \emptyset$. There exists a countable $H \prec_1 V$ such that T, $ST \in H$. Then

$$T_H = T \cap H$$

is countable. So $S(T_H)$ is countable by mild stability of T. Let

$$m : H \longrightarrow m[H]$$

be the Mostowski collapse. Thus $m[T_H] = m(T)$ and $S(T_H)$ is $\{m^{-1}[p] \mid p \in S(m(T))\}$. Hence $S(m(T))$ is countable. And so

$$S(m(T)) \in m[H]$$

by arguments in effective descriptive set theory (similar to showing every countable Δ^1_1 set of hyperarithmetic reals is a member of $L(\omega_1^{CK})$). Then $S(m(T)) \neq \emptyset$ because $m(T)$ is countable. Choose $p \in S(m(T))$. Then $m^{-1}(p) \in ST$.

Lemma 25. *If T is mildly stable, then T is type-complete.*

Mild stability also helps to resolve the question of type-admissibility.

Let A^+ be be the least Σ_1 admissible set with A as a member. Call A **strongly admissible** if $< A, Z \cap A >$ is Σ_1 admissible for all $Z \in A^+$.

Lemma 26. *If T is mildly stable, A is strongly admissible and $T \in A^+$, then T is type-admissible.*

References

[1] Barwise, J., *Admissible Sets and Structures. An Approach to Definability Theory*, Perspectives in Mathematical Logic. Springer-Verlag, Berlin 1975.

[2] Devlin, K. J., *Aspects of Constructibility*, Lecture Notes in Mathematics. Springer-Verlag, Berlin 1973.

[3] Knight, Julia F., *Prime and Atomic Models*, Jour. Symb. Logic **43** (1978), no 3, 385-393.
[4] Nadel, M., *$\mathcal{L}_{\omega_1,\omega}$ and Admissible Fragments*, In: *Model-Theoretic Logics*, Edited by J. Barwise & S. Feferman, Perspectives in Mathematical Logic. Springer-Verlag, Berlin 1985, 271-316.
[5] Sacks, G. E., *Models of Long Sentences II*, forthcoming.

A UNIVERSALLY FREE MODAL LOGIC

Syraya Chin-mu Yang
Department of Philosophy
National Taiwan University, 1, Sec. 4, Roosevelt Road
Taipei 10617, Taiwan
cmyang@ntu.edu.tw

The main burden of this paper is to present a universally free modal logic, quantified S5 modal system in character with a universally free logic as its underlying system. I propose that to give a heuristic characterization of the metaphysical notion of modality, typically necessity and possibility, in terms of the axioms of a normal quantified modal system with S5-models in possible world semantics, two metaphysical presuppositions are called for:

(i) There could be nothing at all;
(ii) There are contingent objects and an object may exist in distinct possible worlds.[1]

In view of the first presupposition, it is striking that in constructing appropriate possible world semantics, the frame of possible worlds under investigation should include possible worlds with the empty domain. And a much more straightforward way to express the second presupposition explicitly in the intended semantics is to stipulate that the language in use must include a collection of names, taken as constants, and quantifying-in *de re* constructions. The aforementioned considerations suggest that the desired modal system should take a certain version of universally free logic

[1]It is not my intention here to give a conceptual analysis of the metaphysical concepts of necessity and possibility. Nor will I propose a certain criterion, or some criteria, to classify what counts as a necessary truth (or possible truth, respectively). My aim is to apply the Tarskian characterization of the semantic conception of truth (with regard to first order logic) to a certain normal quantified modal system so that modal truths can be characterized by virtue of a classification of true modal sentences (i.e., sentences containing modal operators) based on the intended possible world semantics appropriate for the constructed modal system. I hope this heuristic approach will shed light on an illumination of the metaphysical concepts of necessity and possibility.

as its underlying system. However, the acceptance of possible worlds with empty domain and contingent objects in the ontology under investigation would give rise to certain notorious semantic harassments if we stick to the Tarskian-objectual-semantics-based possible world semantics, that is, the unintelligibility of modal sentences with names as rigid designators (e.g. '$\Box Fa$') and/or *de re* constructions (e.g. '$\forall x \Box Fx$'). For once names are interpreted as rigid designators, some of them may have no referents in some possible worlds; they are empty names in those worlds, so to speak. It follows that we would be in no position to determine the truth value of 'Fa' in a world where 'a' is an empty name, let alone the truth value of '$\Box Fa$'. A similar problem occurs with *de re* constructions. A unifying treatment of the legitimacy and intelligibility of sentences containing *de re* constructions in a quantified modal system is required.

I shall start with a specification of some presuppositions for an acceptable notion of metaphysical modality to illustrate that if a desired quantified modal system is to be constructed on the basis of an appropriate possible world semantics, the structure of which is induced by these presuppositions, then the language in use would inevitably contain sentences with names as rigid designators and sentences with quantifying-in *de re* constructions. Clearly, sticking to the Tarskian objectual semantics, the problem concerning the intelligibility of modal sentences with names as rigid designators has its roots in the so-called truth value gaps caused by the acceptance of names for contingent objects. I propose that this problem can be overcome by putting forward a formation rule, which stipulates that no name occurs in a formula except for the contexts "$x = a$", when a indicates the place where a name used as a rigid designator may occur. I further argue that alleged unintelligibility of quantifying-in *de re* constructions has the same origin as the acceptance of names as rigid designators for contingent objects. Accordingly, to secure the legitimacy and intelligibility of quantifying-in *de re* constructions, all that is required, by the same reasoning, is to put forward a formation rule for modal operators, which stipulates that no free variable x appears within the scope of a modal operator, except for the contexts "$y = x$", where y is bound to a certain y-binding universal/existential quantifier. Buttressed with the proposed formation rule for names, I present a system of universally free logic IQ, which includes names as rigid designators. Moreover, taking IQ as the required underlying system, a desired quantified modal system $IQS5$ with the proposed formation rule for modal operators can be constructed. The two extra formation rules will not only legitimate the *de re* constructions in the desired modal system but also guarantee that every sentence has a definite truth-value in every possible world. Finally, I shall adopt a recent proposal in which names can be treated as constant quantifiers, and by virtue of appropriate syntactical rule for names, the rigidity of names can be formulated. A desired quantified modal system with names as constant quantifiers can be constructed by appropriate modification of the proposed formation rules.

1. Some Presuppositions for a Naïve Metaphysical Conception of Modality and *de re* Constructions

As is well-known, a modal system may be obtained from a given logical system such as the propositional calculus, by adding to the language, some modal operators together with formation rules to produce modal formulae/sentences, and then supplementing the underlying system with modal formulae/sentences as axioms together with rules of inference for the given modal operators. The typical modal operator is the primitive symbol '$\Box$'; the usual formation rule for '$\Box$' is: if φ is a formula, so is '$\Box\varphi$'. A huge family of modal systems has been constructed in this way since the 1920's. However, misgivings over the intelligibility of sentences with names within the scope of some modal operators and sentences with quantifying-in *de re* constructions have made a majority of logicians and philosophers reluctant to accept quantified modal systems, which would accept contingent objects.[2]

At present, the most popular semantic treatment for modal operators is perhaps the widely accepted Kripkean possible world semantics, which assumes a collection of possible worlds W, and then claims that

(S$\Box$) $\Box\varphi$ is true in a world w iff φ is true in every possible world w^*.

(Sometimes, a certain accessibility relation among possible worlds is assumed.) Admittedly, the language of the underlying first-order system may contain a category of names serving as rigid designators in the sense that at a given interpretation when an object is assigned to a name as

[2]In logical parlance, it is widely agreed that modal expressions of the form '$\Box\forall x\varphi(x)$', where $\forall x\varphi(x)$ is a non-modal formula containing no names are taken as *de dicto* formulae/sentences; while modal expressions of the form '$\forall x\Box\varphi(x)$' are taken (quantifying-in) *de re* formulae/sentences. But it remains open to dispute whether a modal expression of the form '$\Box\varphi(a)$' in which some name a occurs but no free variable is a *de dicto* sentence or a *de re* one. It seems to me that if we intend to adopt the objectual semantics for the first order language in use, then it would be much more appealing to treat '$\Box\varphi(a)$' as a *de re* one. I shall therefore use a more general phrase '*de re* constructions' to refer to both quantifying-in *de re* formulae/sentences and formulae/sentences with names in the scope of some modal operator.

its semantic value, the name will always take the very same object as its semantic value in every world. Now, when contingent objects are admitted in the domains of possible worlds, a name, say a, may become an empty name in some possible worlds in which the supposed semantic value of a does not exist. In this case, adhering to the Tarskian objectual semantics for the underlying system, we would be in no position to determine the truth value of a sentence of the form $\varphi(a)$ in such a world, and an intelligible interpretation of $\Box\varphi(a)$ becomes problematic.

In view of the seeming unintelligibility of modal sentences with names as rigid designators, it is a natural inclination to formulate quantified modal systems without names. Be that as it may, a more bizarre problem concerning the intelligibility of quantifying-in *de re* constructions has threatened the legitimacy of a quantified modal system, which accepts contingent objects in the required domains of possible worlds. For the truth condition for a sentence of this kind, say $\forall x\Box\varphi(x)$, requires that whatever is assigned as the semantic value of x in the actual world must satisfy the predicate $\varphi(x)$ in every possible world. And again, we are in no position to determine if $\varphi(x)$ is satisfied in a world in which the semantic value of x does not exist, and *a fortiori*, talk of the satisfaction of the modal predicate $\Box\varphi(x)$ becomes problematic. A possible solution is to get rid of all modal sentences with quantifying- in *de re* constructions, or to reduce them to some other intelligible formulae, say *de dicto* formulae.

Of course, we may construct a certain quantified modal system without names as rigid designators and quantifying-in *de re* constructions. But it is not clear how to construct an appropriate semantics for the established modal system so that the metaphysical presuppositions stated above can be displayed explicitly in the interpreted language. My primary concern in this paper is to argue that the semantic problems caused by the acceptance of empty names and quantifying-in *de re* constructions can be dealt within the syntactic level. I show how this can be done and then construct a modal system with appropriate possible world semantics.

Let us start with some naïve presuppositions for a metaphysical notion of modality:[3]

---

[3]As my main concern in this paper is with a satisfactory treatment of *de re* constructions, I shall not dwell on the discussion over which one, amongst a variety of modal systems, is the most appealing system for a theory of metaphysical modality. Nor would I be bogged down with the dispute over whether these presuppositions are right ones. What I have in mind is merely to show the possibility of constructing a quantified modal system so that modal theorems under the intended interpretation can be entirely consonant with an acceptable notion of metaphysical modality, which can be characterized to a certain

(M1) Necessity and possibility are modalities, which can be ascribed to things;
(M2) There are contingent objects;
(M3) Nothing can be said about the non-existent except for the fact that they do not exist;
(M4) Everything exists but there could be nothing at all.

Let us further assume, again without further ado, that the frame of possible worlds without the accessibility relation (or the equivalence relation is assumed) for the propositional modal system $S5$, known as $S5$-models, is the most appealing model for a theory of metaphysical modality. Now, to capture our desired metaphysical presuppositions, the required possible worlds for the desired quantified modal system would meet the following conditions:

(P1) A transworld identity relation holds amongst distinct possible worlds, as this will allow us to ascribe a certain possibility or necessity to a given, specified or indefinite, object.
(P2) Some objects may exist in some worlds but may not exist in some others.
(P3) A possible world displays nothing about the non-existent except for the fact that they do not exist.
(P4) There is a possible world, in which nothing exists, i.e., a possible world with the empty domain is included.

For brevity, let us call a class of possible worlds having the aforementioned properties, a Russellian frame (of possible worlds). In view of (P1), I suggest that a naïve and intuitive way to express transworld identity relation amongst possible worlds for a given specified object so that the ascription of a certain possibility or necessity to a certain specified object can be formulated explicitly is to appeal to the rigidity of names in use. The use of 'rigid designator' indicates that once a name has a certain object as the fixed interpretation in a certain world, typically the actual world, it will take the very object as the supposed semantic value in all other worlds. Accordingly, the language in use will contain a category of primitive symbols $a, b, c, \ldots$ (taken as names). And with (P2), (P3) and (P4), when the sup-

extent by the listed presuppositions. Of course, some might have a different thought about a satisfactory notion of metaphysical modality, which may render some distinct semantic requirements. But I hope that my proposed treatment in what follows will be still available for their systems, provided that rigid designators and *de re* constructions are involved.

posed semantic value of a name does not exist in a world, the name under investigation becomes an empty name in that world. Now, for a quantified modal system to accept empty names and empty domains, the underlying system must be a universally free logical system in character, i.e. a system of free logic and also a system of inclusive logic. If the truth conditions of sentences of the underlying system were to be characterized in terms of the Tarskian notion of satisfaction, we would inevitably encounter the notorious problem of truth-value gaps caused by the use of names as rigid designators, which may become empty names. The presuppositions (M3) and (P3) indicate that there might be no satisfactory interpretation for a sentence containing empty names on the classical objectual semantics; and the collapses of the standard possible world semantics would follow immediately. For given that the sentence $\varphi(a)$ may be truth-valueless in some possible worlds, how could we determine the truth value of the modal sentence $\Box\varphi(a)$?

Analogously, quantifying-in *de re* constructions are required as long as we wish to formulate the ascription of a certain possibility or necessity to unspecified/indefinite objects. On the Tarskian objectual interpretation of quantification and the standard Kripkean possible world semantics, the truth condition of a *de re* sentence of the form '$\exists x\Box\varphi(x)$' requires that there is an object in the actual world, taken as the semantic value of x, which satisfies the predicate $\varphi(x)$ in every world. More specifically, when a modal operator is in the scope of an x-binding quantifier with regard to some variable x, and once an individual is assigned to x as its semantic value in the actual world, the variable in question would become a rigid designator. For the occurrences of the same variable within the scope of the given modal operator will take the very same object as its semantic value in the related possible worlds.[4] The supposed rigidity of variables in the scope of a modal operator makes quantifying-in *de re* constructions seemingly unintelligible on the intended semantics based upon the Russellian frame. A number of philosophers have suggested expelling formulae of this kind from a modal system. But are quantifying-in *de re* constructions dispensable? Intuitively, granted the rigidity of variables in the scope of a modal operator, a quantifying-in *de re* sentence can be used to express whatever happens to the very same unspecified object in related worlds. We can find

[4]Bostock ((1988), p.347) has rightly remarked that "a single interpretation will treat every occurrence of the variable that is bound to the same quantifier as designating the same object. Hence, if some relevant occurrences of the variable are in the scope of an embedded modal operator, a particular interpretation of that variable will treat it as designating the same object at all further worlds introduced by the embedded operator."

that without quantifying-in *de re* constructions, the transworld identity of unspecified/indefinite objects could not be formulated in a first order language. That is, the rigidity of variables occurring in *de re* contexts enables us to talk about the very same object, though indefinite, without the use of a name. And no *de dicto* constructions can ever achieve such a function. In ascribing metaphysical modality to indefinite objects, quantifying-in *de re* constructions are indispensable. For example, there would be no appropriate *de dicto* rendering of the ordinary modal statement,

(1) Everything, which is actually an F, is necessarily a G.

But, admittedly, talk of modal sentences containing quantifying-in *de re* constructions, e.g. $\exists x\Box\varphi(x)$, would become unintelligible if (M2) and (M4), hence (P2) and (P4) are taken into account. For the acceptance of contingent objects would imply that a given object assigned as the semantic value of some variable x may not exist in some worlds, say w_1, such that we are in no position to determine if $\varphi(x)$ is satisfied in w_1. Accordingly, the determination of the truth value of $\forall x\Box\varphi(x)$ would be rather problematic. The core of the issue is this: Are quantifying-in *de re* constructions really unintelligible? If not, what else could we do about them? For those who accept (M1)-(M4) as presuppositions for an acceptable notion of metaphysical modality, making *de re* constructions intelligible remains first and foremost a philosophical endeavor.

2. A Syntactic Treatment of *de re* Constructions

During the last few decades, several attempts at a satisfactory semantic treatment of sentences with empty names have been proposed, such as the appeal to three-valued semantics, semantics with inner/outer domains, free semantics based on the notion of supervaluation, or free semantics based on the principle of falsehood. Some have tried to adopt the substitutional interpretation of quantification, instead of the Tarskian objectual semantics. Nonetheless, none of them looks promising. It is not my intention here to argue the pros and cons for each of them. Instead, I shall propose a syntactic treatment of the legitimacy and intelligibility of sentences with names as rigid designators and later I shall show how to extend this treatment to sentences containing quantifying-in *de re* constructions in a desired quantified modal system.

The attempt to deal with truth value gaps caused by the use of empty names in the syntactic level is by no means brand-new. Indeed, this approach has its origin in the Russell-Quine's dismissal of names, according to which one can always do without names in a logical theory without losing the ontological commitment of the intended theory. Quine ((1951), vi) claimed that this is "a way of simplifying the theory and also helping to clarify ontological considerations." Nonetheless, a logical theory without names may lack the required denotation mechanism for the rigidity of names for specified objects, which will allow us to denote the very same specified object in every possible world. Of course, we may, as Quine suggested, be able to specify or pick up a unique object, to which a certain name in ordinary discourse such as 'Socrates' is supposed to refer, by introducing a predicate, say 'being identical with Socrates', or 'Socratizes', and then stipulate its extension in such a way that the predicate in question could be true solely of the very object under investigation. However, such a replacement for names appears not only redundant but also awkward. Things would become worse when modal sentences are taken into account. It is striking that the desired transworld identity relation can be expressed explicitly by virtue of the rigidity of names. This allows us to talk about the possibility or necessity of the very same specified object. This is a function that the substitution of predicates for names, no matter how appropriate it may be, could not supply.

One may find that not all sentences containing empty names with regard to a world cannot have a determinate truth-value in that world. In view of (M3) and (P3), if we adopt a modified strong sense of the rigidity of names, according to which a name takes the very same object as its semantic value in every possible world, then all existential assertions, even though names are involved, have a definite truth value in every possible world because every possible world has a fixed domain. Bearing this in mind, some free logicians have suggested that in order to ensure the legitimacy of sentences with names, say Fa, all that is required is the addition of the related existential assertions, say '$\exists x = a$' or '$E!a$' (read as 'a exists'). With the help of a sentence of this form, we can legitimately derive Fa by an application of *Instantiation* to $\forall x Fx$ together with the additional premises '$\exists x = a$' or '$E!a$'.

It is undisputable that a sentence with names would have a definite truth value in a world in which all related existential assertions are true. But, when any of these existential assertions is false, the notorious truth-value gaps would rear their ugly head again. For given that "$\exists x = a$" is false in a world w, the truth value of Fa in w become problematic again. This indicates that adding extra existential assertions would not suffice to ensure that every sentence with names has a truth-value in every world. Some more promising and effective approach is called for.

Although the introduction of extra existential assertions cannot remove all truth-valueless sentences from a system of free logic, this approach does shed a light on the search for a more effective way to deal with the problem of truth-value gaps caused by the acceptance of empty names. Bearing in mind that all existential assertions and their cognates, though containing names, still have a definite truth value in every world, as (P3) suggests, we can keep names in the language in use with their occurrences confined exclusively to the contexts of this kind. More specifically, names can only occur in existential assertions. For brevity, let us assume that the language in use contains only one type of expressions for existential assertions, namely formulae of the form '$x = \tau$' where τ is a term and x is a variable which is already bound to an x-binding universal/existential quantifier, and nothing else.[5] Clearly, any sentence of the form either $\forall x x = a$ or $\exists x x = a$ has a definite truth value in every world regardless the existence of the denotation of the name a. Accordingly, to ensure that no truth value gaps caused by the acceptance of empty names appear, all that is required is to put forward a formation rule for names, which stipulates that a name a can only occur exclusively in the contexts '$x = a$'. With this restriction, truth value gaps caused by the use of names as rigid designator disappear automatically. Moreover, in a desired quantified modal system with this formation rule, all *de re* sentences of the form $\Box\varphi(a_1, \ldots, a_n)$ will have a definite truth value in every world.

We still need to deal with the intelligibility of quantifying-in *de re* as far as a desired quantified modal system is concerned. In the last few decades several attempts have been proposed. A simple solution, as Kit Fine and David Kaplan have suggested, is to reduce each *de re* construction to a logically equivalent *de dicto* counterpart. However, there are no effective methods by virtue of which we can rephrase every *de re* sentence as a *de dicto* one, nor can each *de re* sentence be reduced to a *de dicto* one. (See Hughes and Cresswell (1984), p.68). Truly one can introduce into a modal system extra axioms so as to reduce, or to translate, some *de re* sentences into *de dicto* ones, such as the well-known Barcan formula and its

[5]Note that a natural language may by and large contain a variety of phrases to express existential assertions. For example, in English we have verbs such as 'exist', 'lives', ... etc., and adjectives 'existent', 'non-existent', 'identical with', and the like; while a number of free logicians prefer to add to the language in use a logical constant '$E!$' to form sentences of the form '$E!\vartheta$', for any term ϑ, read as "ϑ exists". It is observed that in a language with identity symbol '=', such a constant is superfluous because on the objectual interpretation of quantification, the existential assertion '$E!\vartheta$' can be contextually defined in terms of '$\exists x x = \vartheta$.'"

converse, *viz.*,

(BF) $\forall x \Box \varphi(x) \rightarrow \Box \forall x \varphi(x)$
(CBF) $\Box \forall x \varphi(x) \rightarrow \forall x \Box \varphi(x)$

Yet, the ontological assumptions required for an appropriate semantics for (BF) are not quite consonant with our naïve notion of metaphysical modality in requiring that everything exists in every world. That is, all worlds have exactly the same domain. But this would contradict our presupposition (M2) and (M4). Moreover, syntactically, *de re* constructions are by no means mere reformulations of certain *de dicto* counterparts with a pseudo form; they resist being reduced to corresponding *de dicto* counterparts. Interestingly some quantifying-in *de re* sentences can be comprehensively interpreted on the Russellian frame, for example,

(2) $\forall x \Box (\exists y y = x \vee \neg \exists y y = x)$.

But why does an intelligible interpretation appear to be unattainable for quantifying-in *de re* sentences in general? Quine rightly pointed out that the unintelligibility of *de re* sentences in general has its roots in the inapplicability of objectual satisfaction. Nonetheless, inapplicability does not lie upon the failure of substitutions, as Quine strongly argued. Rather, it stems from the acceptance of contingent objects. As noted, to make a quantifying-in *de re* sentence of the form '$\exists x \Box Fx$' true, there should exist an object in the actual world such that the very same object satisfies the given predicate within the scope of the given modal operator in every world. Since the Russellian frame accepts contingent objects, an object (in the actual world) assigned to a variable x as its semantic value (with regard to the actual world) may not exist in some other worlds so that in such a distinct world, we are in no position to determine whether the very object would satisfy the associated predicate(s) within the scope of a modal operator which is in turn within the scope of a certain x-binding quantifier. Thus, the root of the unintelligibility of quantifying-in *de re* sentences is precisely the same as that of the truth value gaps caused by the use of empty names. That is to say, the difficulty with the application of objectual satisfaction to quantifying-in *de re* constructions is precisely the same as the unintelligibility of applying the notion of objectual satisfaction to sentences with empty names. Therefore, the basic idea which leads us to a radical solution, in the syntactic level, for the truth-value gaps caused by the use of empty names also lends itself to dealing with the problem of quantifying-in *de re* constructions. More specifically, just as our restriction on the use of a name a to contexts '$x = a$' can avoid the occurrences of truth-value gaps caused by the use of empty names, we can put forth a formation rule for modal operators to stipulate that, apart from the contexts '$y = x$', there are no free occurrences of a variable x within the scope of a modal operator. With this restriction, the language in use will contain certain quantifying-in *de re* constructions which can be intelligibly interpreted on the Russellian frame.

To sum up, in order to guarantee that the notion of objectual satisfaction is applicable to *de re* constructions in a quantified modal system, we need: a formation rule for the use of names to stipulate that a name a can only occur in the contexts '$y = a$'; and a second formation rule to stipulate that no free variable x occurs within the scope of a modal operator, except for the contexts '$y = x$', where y is a bound occurrence of some distinct variable within the scope of the given modal operator. With these two formation rules, the language in use can include names as rigid designators and quantifying-in *de re* constructions but every sentence can be intelligibly interpreted on the Tarskian objectual semantics for the underlying system and the standard Kripkean possible world semantics for the desired quantified modal system. The two persistent problems with the legitimacy and intelligibility of *de re* constructions in a quantified modal system can be therefore solved in quite a similar way. In what follows I shall show that these treatments are formally adequate by constructing a desired quantified modal system. But first, the required underlying system is in order.

3. The Underlying System *IQ*: A System of Universally Free Logic with Rigid Designators

We are now ready to present a required underlying system, a universally free logic in character, referred to as the system *IQ*.

(A) The language

The language of *IQ* has an alphabet including the following categories of primitive symbols: name letters $a, b, c, \ldots$, variables $x, y, z, \ldots$, predicate letters $F, G, H, P, \ldots$, the identity symbol '$=$', logical operators $\neg, \rightarrow, \forall$, and auxiliary symbols '(', ')'. All names and variables are terms; we use the Greek lowercase letters τ, ι, $\tau_1, \tau_2, \ldots$, as symbols for terms in general. (Other usual logical operators $\wedge$, $\vee$ and $\exists$ can be introduced in the usual way.).

Atomic formulae The atomic formulae of *IQ* are strings of the form given by the following formation rules:

(i) For any variable x, any term τ, '$x = \tau$' is an atomic formula;

(ii) For any n-place predicate P other than '$=$' and for any variables $x_1, \ldots, x_n$, '$Px_1 \ldots x_n$' is an atomic formula.

Formulae The formulae of IQ are defined as follows:

(iii) An atomic formula is a formula;
(iv) If φ and ψ are formulae, so are φ, $(\varphi \to \psi)$ and $\forall x\varphi$.
(v) A sequence of symbols counts as a formula only if it can be shown by a finite number of applications of (i)-(iv).

The formation rules (i) and (ii) together stipulate that a name a can only occur in the contexts '$x = a$'. To make the desired modal system more natural, I assume the underlying system deals exclusively with *sensible formulae*, namely formulae of the type wherein neither vacuous quantifiers nor multiple bondages of quantifiers appear.[6]

(B) The basic semantics

Our intended semantics will be established on the basis of an S5-model of the Russellian frame of possible worlds, which meet the requirements (P1) – (P4). A Russellian frame of possible worlds is assumed. More specifically, a Russellian frame is a complex $\langle \mathcal{W}, \mathcal{D}, \mathcal{I} \rangle$, where $\mathcal{W}$ is a non-empty set of worlds, $\mathcal{D}$ is a function assigning a set D_w to each world w in $\mathcal{W}$, and $\mathcal{I}$ is a function assigning an interpretation function I_w to each world w in W. To be more precise, the semantic structure of a world w can be understood as an ordered pair $\langle D_w, I_w \rangle$, where D_w, taken as the required domain of w, is the collection of objects existing in w and I_w is the proposed interpretation of the language in use in the world w.[7] In particular, I_w will assign a unique object, say $\mathbf{a}_i$, from the class of all objects existing in all possible worlds (i.e. $\bigcup_{w \in \mathcal{W}} D_w$) to each name a_i as its semantic value (i.e. its denotation) in w, and an n-ary relation on D_w to each n-place predicate letter.[8] To express the required transworld identity via the rigidity of names, we stipulate that

(RN) $I_w(a) = I_u(a)$, for each name a, and for each w and u in $\mathcal{W}$.

Note that $I_w(a)$ may not be in the domain D_w since once an object, say $\mathbf{a}$, in a given world is assigned to a name a as its semantic value in that world, the very object will serve as its semantic value in all other worlds.

---

[6]The term 'sensible formula' used in this sense was proposed by David Bostock in his lectures on Elements of Deductive Logic, Hilary Term, 1993, Oxford University.

[7]Since contingent objects are accepted and since worlds with the empty domain are included, domains of Russellian worlds need not meet either of the following conditions:

(a) $D_w = D_u$ for every pair of worlds w and u in the Russellian frame;
(b) $D_w \neq \emptyset$ for any world w in the Russellian frame.

[8]For the sake of simplicity, here I dispel the possibility of assigning to a name a purely merely possible object, an object which does not exist in any possible world. Nor would I insist that only actual things (i.e. objects existing in the actual world) can have a name.

And for each non-empty world w, I_w will assign to each n-place predicate a set of n-tuple of objects in D_w. Given an established I_w over names and predicate letters, the truth-value of sentences containing no empty names in a world (with a non-empty domain) can be fully determined by the standard Tarskian objectual semantic treatment (e.g. Chang and Keisler (1991)). The determination of the truth-value of a sentence containing some empty names would cause no problem in our proposed semantics because an empty name a_i can only occur in the context of '$x = a_i$' where the variable x is in the scope of some quantifier. Now given that the proposed semantic value, $\mathbf{a}_i$, of a_i does not exist in the world w under investigation, '$x = a_i$' would be false in w under any assignment of the variables in use. This will dismiss the threat of truth value gaps caused by the use of empty name. What remains is to show how to determine the truth-values of sentences in a possible world with the empty domain, referred to as $w_\emptyset$. Recall the presupposition (M3) – Nothing happens to the non-existent except for the fact that they do not exist, a possible world with the empty domain will display nothing about anything except for the fact that none of them exists. It follows that no non-trivial interpretation of predicate letters is available in $w_\emptyset$. The best we can do is to trivially assign the empty set to each predicate letter as its semantic value in $w_\emptyset$. It would make no sense to ask whether a given interpretation of names, or any assignment of variables, would satisfy a formula, say 'Fx', in $w_\emptyset$, when every predicate letter is assigned with the empty set as their extensions. However, a widely accepted convention is available, as far as the determination of truth-values of sentences in $w_\emptyset$ is concerned. That is, a sentence beginning with an existential quantifier would count as not true in $w_\emptyset$, as nothing can be said to satisfy the formula that lies immediately within the scope of the given existential quantifier. By contrast, by the interdefinability of quantifiers, a sentence beginning with a universal quantifier will be true in $w_\emptyset$.

(C) Axiom-schemata and rules of inference

A completed set of axioms for IQ is characterized by the following axiom-schemata:

(A1) $\forall\mathbf{x}\varphi$, for φ a CPC-tautologous formula.

(A2) $\forall\mathbf{x}(\forall x(\varphi \to \psi) \to (\forall x\varphi \to \forall x\psi))$, if x occurs freely in both φ and ψ.

(A3) $\forall\mathbf{x}(\forall x(\varphi \to \psi) \to (\varphi \to \forall x\psi))$, if x occurs freely in ψ but does not in φ.

(A4) $\forall x\neg\forall\neg y = x$, where y is distinct from x.

(A5) $\forall \mathbf{x}(y = \tau \to (\varphi \to \varphi(y/\tau)))$, where τ is a term and $\varphi(y/\tau)$ results from φ by substituting τ for y, if it is syntactically legitimate, in zero or more places in φ.

And the only rule of inference required for IQ is the familiar (MP):

(MP) From φ and $(\varphi \to \psi)$, ψ can be derived.

The expression $\forall x \varphi$' is to stand for the closure of φ; and by a CPC-tautologous formula is meant a formula of the form precisely like a tautology in classical propositional calculus. The notions of a *derivation of a formula*, a *derivation of a formula from a set of assumptions*, and a *theorem* will be defined in the usual way.

This completes the formalization of the universally free logic IQ. The soundness theorem, i.e. every sensible sentence which is a theorem of IQ is valid on the Russellian frame, can be easily proved. For there is an easy routine to check that all instances of (A1)-(A5) hold in every world; and as Quine in his first edition of *Mathematical Logic*, (also, Hailperin (1953)) has shown, (MP) preserves validity in a first-order system in which no theorems are open formulae. Moreover, adopting Hailperin's ideas and approach (Hailperin (1953)), the completeness of IQ can be proved.[9]

There are some interesting connections between IQ and other established system of free logic, for example, van Fraassen's IF (1966), Mayer and Lambert's FQ (1968) and Leblanc's LQ (1968). Van Fraassen's IF results from Quine's IML (Quine, (1954)) by adding to IML the two usual axioms for identity. If we dismiss names from the language in use, then IQ is substantially a subsystem of IF. Meyer and Lambert's FQ includes, in addition to our (A1)-(A3) and (MP), three other axiom-schemata, i.e. (Gen), (FQ1) and '$\forall x E!x$'(which asserts that everything exists, and hence essentially equivalent to our (A4)). Interestingly, FQ is a system without identity and it adopts the Hailperin-Quinean semantic treatment for vacuous quantifiers in the empty domains in that the truth value of a sentence resulting by putting any vacuous quantifier in front of a given sentence would remain the same as the original one. Leblanc suggests that if we add to FQ the identity predicate, and if we adopt Mostowski's (1951) semantic treatment for vacuous quantifiers in the empty domain, we can then get a universally free logic LQ which should be reckoned, as it seemed to him, as the "most satisfactory free logic (without individual constants) to have been proposed yet". More specifically, LQ contains in addition to our (A1),

[9] I have carried out the full semantics and a formal proof of the completeness of IQ with all the details, which turns out to be surprisingly lengthy.

(A4), and (A5), the following three axiom-schemata:

(ML2*) $\forall\mathbf{x}(\forall x(\varphi \to \psi) \to (\forall x\varphi \to \forall x\psi))$, where x occurs freely in ψ,
(ML3*) $\forall\mathbf{x}(\forall x(\varphi \leftrightarrow \psi))$, if x does not occur in φ,
(A8) $x = x$,

and two rules of inference (MP) and (Gen). (ML2*) substantially corresponds to our (A2), except that it also accepts $\forall x\varphi$ even when there is no free occurrence of x in φ; while (ML3*) and (Gen) are superfluous as long as only sensible sentences are taken into account. In the meantime, the closure of (A8), namely, $\forall xx = x$ is derivable in IQ. Thus IQ has a quite intimate kinship with LQ. Nonetheless, IQ differs from LQ in a rather significant way: it accepts names as rigid designators. In particular, when IQ is to serve as the underlying system of a desired quantified modal system, the use of names as rigid designators for contingent objects will enable us to talk legitimately about the ascription of possibility/necessity to specified objects without being subject to the seemingly unintelligible interpretation of *de re* sentences of the form $\Box\varphi(a_1, \ldots, a_n)$. With the acceptance of names and in view of the ontological imports of the construction of the Russellian frame, the system IQ is more satisfactory and natural than other existent systems of universally free logic.

4. A Quantified Modal System with Rigid Designators: A Natural Modal System $IQS5$

Taking IQ as the required underlying system, and with the help of the proposed treatment for quantifying-in *de re* constructions, we can further formalize a quantified modal system, $IQS5$.

(A) The language

The language in use is an expansion of the language for IQ by adding to its alphabet a modal operator '$\Box$', and an extra formation rule for '$\Box$':

($\Box$) If φ is a formula, in which there are no free occurrences of variables apart from the contexts '$= x$', then '$\Box\varphi$' is a formula.

The usual possibility-modal operator '$\Diamond$' can be introduced into the language in use, and the proposed formation rule for '$\Box$' is applied to the modal operator '$\Diamond$' as well. The notion of the scope of a modal operator is to be defined in the usual way.

(B) The semantics

The semantics for $IQS5$ will be established on the basis of S5-models of the Russellian frame of possible worlds, and will inherit the basis semantics for the underlying system IQ. Therefore, we require some appropriate semantic rules for '$\Box$'. Two cases are considered.

Case 1 φ is a sentence. We would then have a simple semantic rule for '$\Box\varphi$':

($\Box s$) When φ is itself a sentence, for any world w, $w \models \Box\varphi$ iff $w^* \models \varphi$ for every world w^*.

Case 2 φ is an open formula with free occurrences of variables amongst $x_1, \ldots, x_n$.

When a modal operator is in the scope of an x-binding quantifier with regard to the variable x, some free occurrences of which are within the scope of the given modal operator, and once an individual is assigned to x as its semantic value in a given world, the very variable would become a rigid designator in that the occurrences of the same variable within the scope of the given modal operator will take the very same object as its semantic value in all related possible worlds. Now, consider the formula '$\Box Fx$', for F, a one-place predicate letter. In speaking of the satisfaction of '$\Box Fx$' under a certain assignment ζ in a world w, it is required that once ζ assigns a certain object in w, say $\mathbf{a}(x)$ as the semantic value of x in w, every free occurrences of x within the scope of the modal operator will take $\mathbf{a}(x)$ as its proposed semantic value in each related possible world w^*, regardless whether $\mathbf{a}(x)$ exists in w^* or does not. For the sake of simplicity, I shall use the notation '$[\mathbf{a}(x_1), \ldots, \mathbf{a}(x_n)]$' to stand for an assignment of variables $x_1, \ldots, x_n$ having some free occurrences in a formula under investigation. This suffices to show that when φ is an open formula with free occurrences of variables amongst $x_1, \ldots, x_n$, then to claim that the formula '$\Box\varphi$' is satisfied by a certain n-tuple at the assignment $[\mathbf{a}(x_1), \ldots, \mathbf{a}(x_n)]$, in a world w, it is required that each variable x_i (for $1 \leq i \leq n$) take $\mathbf{a}(x_i)$ as its semantic value in every related possible world w^*. More generally, we have the following semantic rule:

($\Box f$) When φ is an open formula with free occurrences of variables amongst $x_1, \ldots, x_n$, then for any world w, $w \models \Box\varphi[\mathbf{a}(x_1), \ldots, \mathbf{a}(x_n)]$ iff $w^* \models \varphi[\mathbf{a}(x_1), \ldots, \mathbf{a}(x_n)]$ for every non-empty world w^*, where $[\mathbf{a}(x_1), \ldots, \mathbf{a}(x_n)]$ is to stand for a specified assignment which assigns $\mathbf{a}(x_1), \ldots, \mathbf{a}(x_n)$ to $x_1, \ldots, x_n$ as their semantic values respectively in w and each related possible world w^*.[10]

[10]When we consider the satisfaction of a given modal formula with free occurrences of variables, say $\Box\varphi(x_1, \ldots, x_n)$, at a given assignment in a world, say w, (i.e. $w \models \Box\varphi[\mathbf{a}(x_1), \ldots, \mathbf{a}(x_n)]$, we may assume that $\mathbf{a}(x_1), \ldots, \mathbf{a}(x_n)$ are objects existing in w but some of them may not exist in some related possible worlds u. But notice that in speaking

The characterization of 'a sentence φ's being true in w', and that of the notion of validity will be defined in the usual way.

According to the proposed semantic rules, sentences with quantifying-in *de re* constructions can be legitimately interpreted. For example, suppose that φ is a non-modal open formula with free occurrences of x_i and some bound variables amongst $x_1, \ldots, x_{i-1}, x_{i+1}, \ldots, x_n$. Then clearly, $w_\emptyset \models \forall x_i \Box\varphi$, by our convention for the truth value of sentences prefixed with a universal quantifier in possible worlds with the empty domain. And for every non-empty world w,

(†) $w \models \forall x_i \Box\varphi$ iff $w \models \Box\varphi[\mathbf{a}(x_1), \ldots, \mathbf{a}(x_{i-1}), \mathbf{a}(x_i), \mathbf{a}(x_{i+1}), \ldots, \mathbf{a}(x_n)]$, for every object $\mathbf{a}(x_i)$ in D_w, taken as the semantic value of x_i in w at each assignment, and

$$w \models \Box\varphi[\mathbf{a}(x_1), \ldots, \mathbf{a}(x_{i-1}), \mathbf{a}(x_i), \mathbf{a}(x_{i+1}), \ldots, \mathbf{a}(x_n)],$$

for some specified $\mathbf{a}(x_i)$ in D_w, taken as the value of x_i in w at a certain specified assignment, iff

$$w^* \models \varphi[\mathbf{a}(x_1), \ldots, \mathbf{a}(x_{i-1}), \mathbf{a}(x_i), \mathbf{a}(x_{i+1}), \ldots, \mathbf{a}(x_n)],$$

where $[\mathbf{a}(x_1), \ldots, \mathbf{a}(x_{i-1}), \mathbf{a}(x_i), \mathbf{a}(x_{i+1}), \ldots, \mathbf{a}(x_n)]$ is to stand for some or all related assignment(s) at which $\mathbf{a}(x_1), \ldots, \mathbf{a}(x_{i-1})$, $\mathbf{a}(x_i), \mathbf{a}(x_{i+1}), \ldots, \mathbf{a}(x_n)$ are proposed semantic values of $x_1, \ldots,$ $x_{i-1}, x_i, x_{i+1}, x_n$, respectively. (Here '$\mathbf{a}(x_i)$ in w' and '$\mathbf{a}(x_i)$ in w^*' indicate that x_i will be assigned the same semantic value in two distinct worlds w and w^*).

Together with the formation rule for the use of names as rigid designators, we can then have a modal system in which all *de re* sentences have a definite truth value in any Russellian world. This completes the semantics for the system $IQS5$.

(C) Axiom-schemata and rules of inference

Given the construction of the Russellian frame, which presupposes the construction of S5-models for the propositional modal logic $S5$, the system $IQS5$ includes the following axiom-schemata:

(A1) Every theorem of IQ is an axiom;
(A2) A closure of S5-tautologous formula is an axiom.

The rules of inference are *Modus Ponens* (*MP*), *Necessitation* (*Nec*) and *Genralization* (*Gen*).

---

of the interpretation of names in a world w, say $w \models \varphi(a_1, \ldots, a_n)[\mathbf{a}(a_1), \ldots, \mathbf{a}(a_n)]$, some $\mathbf{a}(a_i)$ may not exist in w.

(MP) From $(\varphi \to \psi)$ and φ, ψ follows.
(Nec) If φ is a theorem, so is $\Box\varphi$.
(Gen) If $\varphi(a)$ is a theorem, where a is a name, so is $\forall x\varphi$(x) in which all occurrences of a in φ are replaced by free occurrences of x.

This needs a further explanation. Clearly, by taking IQ as its underlying system, $IQS5$ preserves all axiom-schemata and the rule of inference in IQ. And the construction of the Russellian frame presupposes the S5-models for prepositional modal system S5, which contains, in addition to all tautologies of the classical prepositional calculus and (Nec) - the rule of inference for modal operator '$\Box$', the following three characteristic axiom-schemata:

(K) $\Box(\varphi \to \psi) \to (\Box\varphi \to \Box\psi)$.
(T) $\Box\varphi \to \varphi$.
(E) $\Diamond\varphi \to \Box\Diamond\varphi$.

It is therefore a natural inclination to take (K), (T) and (E) as the required axiom-schemata. Patently, by applying a certain specified substitution scheme between sentences of the language of IQ and propositional letters in a modal theorem of the propositional modal system S5, the resulting modal sentence would be still valid on the Russellian frame. Imitating the definition of CPC-tautologous formulae, we can call a modal sentence obtained in this way an S5-tautologous sentence, and we may then claim that:

(A6*) An S5-taulogous sentence is a theorem.

It is striking that all instances of (A6*) are *de dicto* sentences. But intuitively, some modal sentences containing quantifying-in *de re* constructions would qualify as theorems as they appear to be valid on the Russellian frame, such as

(3) $\forall x(\Box\exists yy = x \to \exists yy = x)$.

In view of (A6*), the most intuitive way to obtain *de re* theorems such as (3) is to extend (A6*) so that we can apply suitable substitution scheme to open formulae, and then the closure of a modal formula obtained in this way could count as a theorem. That is, to put forth the following as an axiom-schema:

(A6) A closure of an S5-taulogous formula is a theorem,

where an S5-taulogous formula results by consistently substituting formulae of the language in use for every propositional letters of a modal theorem of S5. Obviously, (A6*) is merely a special case of (A6). Nonetheless, (A6) is not sufficient to generate all seemingly valid modal sentences, such as

(4) $\forall x \Box y(y = x \to (Fy \vee \neg Fy))$.

An appealing approach to get a theorem of this kind is to adopt the converse of the well-known Barcan formula, $\Box \forall x \varphi(x) \to \forall x \Box \varphi(x)$, as an axiom-schema. In our language this is rephrased as:

(CBF*) $\Box \forall x \varphi(x) \to \forall x \Box \forall y(y = x \to \varphi(y))$, where $\varphi(y)$ results from $\varphi(x)$ by substituting y for all occurrences of x.

As $\forall x(Fx \vee \neg Fx)$ and $\Box \forall x(Fx \vee \neg Fx)$ are theorems, by an application of (MP), from an instance of (CBF*), (4) follows. It is thereby tempting to take (CBF*) as a desired axiom-schema. However, this approach is redundant. It suggests that to obtain a quantifying-in *de re* theorem by application of (CBF*), one has to show a theorem of the form '$\Box \forall x \varphi(x)$,' which in turn requires that '$\forall x \varphi(x)$' is a theorem already. In other words, to guarantee that everything in the actual world has such-and-such a necessary property, one has to make sure that everything in every world has such-and-such a property. It is more straightforward to obtain a *de re* theorem such as (4) directly from

(5) $\Box \forall y(y = a \to (Fy \vee \neg Fy))$,

by an application of (*Gen*). For in classical logic, a simple way to get a theorem of the form '$\forall x \varphi(x)$' is to apply (*Gen*) — the rule of *Generalization* — to '$\varphi(a)$', provided that a is a name letter open to interpretation. This indicates that the retaining of (*Gen*) is all that required. This completes our discussion of the modal system $IQS5$. The soundness of IQ and that of the propositional modal system S5 would suffice to show that, on the established possible world semantics on the Russellian frame, all instances of axiom-schemata of $IQS5$ are valid, and the rules of inference are evidently validity preserving. The completeness of the system $IQS5$ is more complicated, and I shall not carry out a formal proof. Since $IQS5$ includes theorems containing names as rigid designators for contingent objects and quantifying-in *de re* constructions, and since the assumed Russellian frame meets the presuppositions (M1) - (M4), the system can be reckoned as the most satisfactory natural theory of metaphysical modality to have been proposed yet.

5. A Modal System with Names as Constant Quantifiers

Perhaps putting forth the proposed formation rule to confine the use of names to the contexts '$= a$' at syntactic level is no more and no less than taking the expression '$= a$' as the predicate 'being identical with something named a', or the like, in disguise. Moreover, the rigidity of names can never be explicitly formulated; the fact that names should be taken as rigid designators merely results from certain *ad hoc* semantic stipulation. Without such a semantic stipulation, there is no way to ensure that we can talk about the very same object, to which the name a refers, in every possible world, in terms of '$\Box\varphi(a)$'. It would then follow that the truth of $\Box\varphi(a)$ could not formulate the ascription of a certain necessity to the very object. In response, the best we can do, as it seems to me, is to appeal to a sentence of the form:

$$(6)\ \forall x(x = a \to \Box\forall y(y = x \to \varphi(y))),$$

or the like. But obviously we cannot exclude the possibility that the name a in a sentence of the form $\Box\varphi(a)$ may not take the same object as its semantic value in every world.

I have recently proposed a neo-Fregean account of the sense of names, according to which names can be construed as quantifiers of a special kind, namely constant quantifiers.[11] I have argued that the primary usage of names is to be associated with predicates; no names occur *per se*. And when a name, say a, is associated with distinct predicates, say F and G, to form sentences Fa and Ga, the truths of both sentences Fa and Ga at a single interpretation indicate that the two predicates are predicative of the same object, *namely*, the semantic value of a. This implies that associating the name a with different predicates can be understood as if the use of the same name is to set forth a quantification over the distinct predicates so that they are predicative of the same object. I call such quantification over predicates *constant quantification* in the sense that once a unique object is to be associated with a name, treated as a quantifier, any variable bound to that name will always take the same object as its semantic value. I have further proposed a formation rule for names used as quantifiers in a first-order language: if $\varphi(x)$ is a formula, so is $a_x\varphi(x)$. With this formulation rule for names used as quantifiers, the rigidity of names can be expressed explicitly. For the truth of a *de re* sentence $a_x\Box\varphi(x)$ will require that the object assigned as the semantic value of a at a certain specified interpretation, satisfies the predicate $\varphi(x)$ in every world. This in turn requires that in

[11] For the details, see Yang (2007).

every world, the object assigned as the semantic value of all free occurrences of x in $\varphi(x)$ is precisely the very object assigned as the semantic value of a because all occurrences of x are bound to the quantifier a. (Here, a can be viewed as essentially an x-binding quantifier in $\varphi(x)$.) One can see that this is precisely what the rigidity of names required. To strengthen the rigidity of names in no-modal contexts, we may further add an axiom:

(‡) $a_x\Box\varphi(x) \leftrightarrow \Box a_x\varphi(x)$.

More formally, to construct a quantified modal system with names as constant quantifiers, the language in use will have the same alphabet as that of the language $IQS5$, except that names will be used as quantifiers and all terms are variables. The atomic formulae, and formation rules for $\neg, \rightarrow, \forall$ and $\Box$ will be defined in the usual way with the proposed restriction on the free occurrences of variables within the scope of modal operators:

(NQ1) No free occurrences of any variable x occur within the scope of a modal operator, except for the context '$y = x$' or '$x = y$', where y is bound to a certain y-binding universal/existential quantifier within the scope of the given modal operator.

The formation rule for the use of names in the language of IQ can be modified so as to fit our proposal that name letters are used as constant quantifiers in the following way:

(NQ2) If $\varphi(x)$ is a formula, then $a_x\varphi(x)$ is a formula, for any constant quantifier a_x, provided that free occurrences of x appear exclusively in the contexts '$y = x$' or '$x = y$' in $\varphi(x)$, where y is bound to a certain y-binding universal/existential quantifier.

The semantic treatment for constant quantifiers is exactly the same as usual: at a specified interpretation, a unique object can be associated with a name a used as a constant quantifier to the extent that whenever a variable is bound to the name in question, all its occurrences will receive the very object as their semantic value. Other semantic treatments will remain untouched, and the axiom-schemata will be the same.

A final remark. By taking names as quantifiers thus formulated, it would be more natural to take formulae/sentences with names occurring within the scope of some modal operator as *de re* constructions if *de re* constructions intend to express the ascription of modality to objects. In particular, a formula/sentence of the form $a_x\Box\varphi(x)$ suggests that it is a quantifying-in *de re* construction in character. Surprisingly, whoever prefers to hold a *de dicto* reading for sentences of the form $\Box a_x\varphi(x)$ may find that with (‡), every alleged *de dicto* sentence of the form $\Box a_x\varphi(x)$ can be reduced to a *de re* one, namely, $a_x\Box\varphi(x)$.[12]

References

[1] Bostock, D., 'Necessary Truth, and a *Prori* Truth', *Mind*, vol. 97 (1988), 343-79.

[2] Chang, C.C., and Keisler, H. J., *Model Theory*, 3rd edition, Amsterdam: North-Holland, 1990.

[3] Hailperin, T., 'Quantification Theory and Empty Individual Domains,' *The Journal of Symbolic Logic*, vol.18(1953), 197-200.

[4] Hughes, G. E., and Cresswell, M. J., *A Companion to Modal Logic*, London: Methuen, 1984.

[5] Leblanc, H., 'On Meyer and Lambert's Quantification Calculus FQ,' *The Journal of Symbolic Logic*, vol.33(1968), 275-80.

[6] Meyer, R. K., and Lambert, K., 'Universally Free Logic and Standard Quantification Theory,' *The Journal of Symbolic Logic*, vol.33(1968), 8-26.

[7] Mostowsti, A., 'On the Rules of Proofs in the Pure Functional Calculus of the First Order,' *The Journal of Symbolic Logic*, vol.16(1951), 107-11.

[8] Quine, W. V., *Mathematical Logic*, revised edition, Cambridge, Mass.: Harvard University Press, 1951.

[9] ———, 'Quantification and the Empty Domain,' *The Journal of Symbolic Logic*, vol.19 (1954), 177-9.

[10] Van Frassen, B., 'The Completeness of Free Logic,' *Zeitschrift für Mathematische Logik und Grundlagen der Mathematik* 12 (1966), 219-34.

[11] Yang, Syraya C. M., 'Proper Names as Quantifiers: A Neo-Fregean Account of the Sense of Names', *EurAmerica* 37 (2007): 183-225.

[12]Part of the research for this paper is sponsored by a grant from National Science Council, Taiwan (No. NSC96-2411-H-002-027-MY3). I wish to express my deep gratitude to an anonymous referee whose comments and suggestions help me to refine some formulations of my ideas, and to clarify several conceptual analyses with regard to some substantial points I propose in this paper.

AUTHOR INDEX